Pulsars

A Series of Books in Astronomy and Astrophysics

EDITORS: *Geoffrey Burbidge*
Margaret Burbidge

Pulsars

Richard N. Manchester
COMMONWEALTH SCIENTIFIC AND
INDUSTRIAL RESEARCH ORGANIZATION

Joseph H. Taylor
UNIVERSITY OF MASSACHUSETTS

W. H. Freeman and Company
SAN FRANCISCO

Library of Congress Cataloging in Publication Data

Manchester, Richard N
   Pulsars.

   (A Series of books in astronomy and astrophysics)
   Bibliography : p.

   Includes index.
   1. Pulsars. I. Taylor, Joseph H., joint author. II. Title.
QB843.P8M36   523   77–4206
ISBN 0–7167–0358–0

Printed in the United States of America

9 8 7 6 5 4 3 2 1

*To Jocelyn Bell, without whose perceptiveness
and persistence we might not yet have had
the pleasure of studying pulsars.*

Contents

Tables

Preface

Few events have had such an impact on the course of astrophysical research as the discovery of pulsars in 1967 by the group of Cambridge astronomers led by Antony Hewish. This fact was recognized by the Swedish Academy of Sciences when it awarded Hewish the 1974 Nobel Prize in physics in recognition of the work that culminated in the discovery of pulsars.

The discovery stimulated the development of new techniques for the study of rapidly varying phenomena over the whole range of the electromagnetic spectrum, from low radio frequencies to the gamma-ray band. Pulsars are nearly ideal as probes of the interstellar medium—they are pulsed, highly polarized, and widely distributed throughout the galaxy—and have been widely utilized as such. On the theoretical side, pulsars have been the stimulus for a great deal of work on the electrodynamics of rotating magnetized stars, coherent radiation processes, and the structure and properties of neutron stars. The identification of pulsars with neutron stars provided the first observational evidence for these extremely compact objects first predicted theoretically by Walter Baade and Fritz Zwicky in 1934. Previous observational searches for neutron stars had been confined to attempts to detect the thermal X-ray emission from neutron star surfaces—as John A. Wheeler has remarked, it was not realized that neutron stars would come equipped with a handle and a bell!

In the few years since their discovery, remarkable progress has been made in the understanding of pulsars. By 1969, the rotating neutron star model for pulsars was well established and most subsequent work has been based on this model. It should be remembered, however, that there may be other explanations for the phenomena that we observe. For example, Mertz (1974) has proposed that the pulses are emitted from mode-locked masers formed by guided waves propagating around white dwarf stars. Nevertheless, up to the present time none of the alternative models has been able to provide as satisfactory an explanation for the wide variety of pulsar phenomena as those built around the rotating neutron star model.

Although much work remains to be done, the basic observational characteristics of pulsars are now rather well established. By contrast, many of the theoretical aspects, particularly the structure of neutron stars and the electrodynamics of pulsar magnetospheres, are still evolving rapidly. No doubt because much of this work is at the frontiers of physics, many different avenues are being explored, often leading to different and in some cases contradictory conclusions. Partly because of this state of affairs and partly because of our own backgrounds, the major part of this book is devoted to a description of the observed properties of pulsars. Discussion of the theoretical models is largely confined to two chapters, albeit the longest chapters in the book.

The book attempts to provide a comprehensive introduction to the subject of pulsars at a level suitable for advanced undergraduate and graduate students. Because of the comparative youthfulness of the subject it has also been possible to provide a reasonably complete review of the state of pulsar research at the present time. We therefore hope that the book will also prove useful to practicing astronomers and astrophysicists.

Of course, most of the information has come from the now very extensive literature on pulsars. But we have also drawn on a fairly large pool of unpublished data, particularly for many of the tables and figures. In general, except for important historical papers, we have cited only the most recent work in a given area—references to earlier material can usually be found in these papers.

We benefited greatly from the advice and suggestions of colleagues who read part or all of the manuscript, namely, J. L. Caswell, W. A. Coles, W. M. Goss, R. N. Henriksen, F. K. Lamb, and B. A. Manchester. We especially thank J. A. Roberts for his careful reading of the entire manuscript. A number of colleagues sent us preprints, previously unpublished data, or illustrations for which we are grateful. For making available the facilities of the CSIRO, Division of Radiophysics, during the preparation of this book we thank Dr. J. P. Wild. In particular, we thank the Publications office for their efforts.

March 1977   *R. N. Manchester*
                                                      J. H. Taylor

Pulsars

General Properties of Pulsars

The Discovery of Pulsars

One of the most remarkable astronomical discoveries in recent memory was the detection in late 1967 of the clocklike radio pulses emitted by objects that have come to be called pulsars. The discovery was made by Jocelyn Bell and Antony Hewish at Cambridge University, and it came about as a direct (but unexpected) result of putting into operation a large radio telescope array designed to study the interplanetary scintillation of compact radio sources. The telescope is a rectangular array containing 2,048 full-wave dipoles operating at 81.5 MHz and covering nearly five acres of land (Figure 1-1). The dipoles are phased to produce several beams, each having a width of about 1° in right ascension and 6° in declination, and separated by 6° in the north–south direction. During the early scintillation surveys, and hence during the initial discovery of pulsars, four beams were used simultaneously with receivers having time constants of about 0.1 s (Hewish and Burnell, 1970). Thus, a large fraction of the sky was under frequent surveillance with both high sensitivity and fast recording techniques, a situation that proved fortuitous for the discovery of pulsars.

1-1 Part of the 81.5 MHz array at the Mullard Radio Astronomy Observatory with which pulsars were discovered. [Courtesy of Dr. A. Hewish.]

Radio telescopes with sensitivity adequate for detecting pulsars had been in existence since the 1950s. However, because rapid time variations in emissions from celestial sources were unknown (except for those within the solar system), receivers and recording devices were generally equipped with time constants of several seconds to smooth random noise fluctuations. The *mean* flux level from most pulsars is quite low, well below the detection threshold of early surveys made with such systems.

The group at Cambridge had great difficulty convincing themselves that the strange, sporadic, signals they were observing had been emitted by naturally occurring astronomical objects. Clearly distinguishable pulses showing periodicity were first recorded on November 28, 1967 (Figure 1-2). During the next eight weeks Hewish and his colleagues systematically eliminated all of the more plausible explanations for the strange signals. Man-made signals transmitted from space probes or reflected from the moon or planets were ruled out because the absence of any parallax greater than about two arc minutes showed that the source lay far outside the solar system. When it was realized that the time duration of the emitted pulses was of the order of 20 ms, and hence (on the basis of

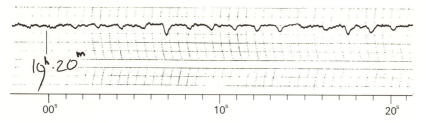

$10^h.20^m$

| 00ˢ | | | | | | | | | 10ˢ | | | | | | | | | 20ˢ |

1-2 The first chart record of individual pulses from a pulsar, PSR 1919 + 21, recorded on November 28, 1967. Increasing intensity is downwards on the chart. [From Hewish, 1975. Copyright American Association for the Advancement of Science.

light travel-time arguments) that the source could not be larger than the earth, the possibility that the signals might have been transmitted by an extraterrestrial civilization was briefly entertained. However, no recognizable code could be discerned in the signals, and the absence of any Doppler variation in the pulse repetition rate (beyond that ascribable to the earth's orbital motion) made it unlikely that the source of emission was located on a planet. When three more similar pulsating sources were detected, it became clear that the sources had to be natural phenomena. The discovery of the first pulsar, PSR 1919 + 21,* was announced by Hewish, Bell, Pilkington, Scott, and Collins on February 24, 1968, in the journal *Nature*.† This paper presented the basic facts and interpretation in a remarkably comprehensive manner, and included a proposal for a model based on either white dwarfs or neutron stars.

The impact of the discovery on the international astronomical community was enormous. Within a few weeks most of the larger radio telescopes in the world were directed toward PSR 1919 + 21, and a torrent of both observational and theoretical papers began to flow into the journals. In 1968 alone more than 100 papers reporting observations of pulsars or their interpretation were published. Most radio observatories made at least a cursory effort to detect additional pulsars. The most concerted efforts took place at observatories with large telescopes of

---

* In the early years of pulsar astronomy it was conventional to name pulsars by a two-letter prefix, e.g., CP, indicating the observatory (Cambridge) and P for pulsar, followed by a four-digit number indicating hours and minutes of right ascension, e.g., 1919. With the advent of more extensive and sensitive pulsar surveys, this system was unable to give unique names to many objects. Because of this, and the desire for a more uniform and readily identifiable nomenclature, the prefix PSR (abbreviation of "pulsar") was adopted for all pulsars. As in the earlier system, the prefix is followed by a four-digit number indicating right ascension (in 1950.0 coordinates), but in addition a sign and two digits indicate degrees of declination. When further resolution is required, declination is given in tenths of a degree by adding another digit.

† For Hewish's own account of the discovery see *Science*, Vol. 188, p. 1079 (1975).

limited steerability, such as the Mills Cross at Molonglo, Australia, the 92-m transit paraboloid at Green Bank, West Virginia, the 305-m spherical reflector at Arecibo, Puerto Rico, and the dipole array at Cambridge. The first four pulsars discovered by the Cambridge group were soon accompanied by a fifth one discovered by a group from Harvard University observing at Green Bank, two more from Cambridge, two from Molonglo, two from Arecibo, and one from Jodrell Bank, England. The first dozen pulsars discovered had broadly similar characteristics, including narrow pulses with variable amplitudes and pulse repetition periods distributed in the range 0.25 to 1.96 s.

Then, in late 1968, the Molonglo group announced the discovery of a much faster pulsar, with a period of only 0.089 s, located near the center of the large supernova remnant Vela X (Large, Vaughan, and Mills, 1968). About the same time, at Green Bank, Staelin and Reifenstein (1968) announced the detection of two sources of pulsed emission near the Crab Nebula, the most widely studied of all supernova remnants. Observations at Arecibo showed that one of these pulsars was located within five arc minutes of the center of the Crab Nebula and that its period was only 0.033 s. Moreover, it was soon discovered that the period was lengthening at the rate of about one part in 2,000 per year. These observations showed that pulsars are related to supernovae, and that they are probably rapidly rotating neutron stars spinning at the observed pulsation frequency, a suggestion made earlier by Gold (1968). The existence of the Crab pulsar also fulfilled predictions made by Wheeler (1966) and Pacini (1967) before the discovery of pulsars, namely, that the energy source in the Crab Nebula could be a rotating neutron star. (Pacini had also discussed the conversion of rotational energy into magnetic-dipole radiation and hence into particle motions.)

During the first year or so of rather hectic observation of pulsars, workers at various observatories naturally concentrated their efforts on those observational problems readily studied with the equipment at hand. In addition to the already-mentioned search for new pulsars, major efforts were directed toward measuring the characteristics of individual pulses and subpulses (at Arecibo); the mean pulse profiles, pulse timing, and polarization (Jodrell Bank and Green Bank); the radio-frequency spectra (Parkes, Australia); and detailed observations at much higher frequencies (2,300 MHz, at the Goldstone Deep-Space Station, California). There were also observations at optical and shorter wavelengths. As soon as positions of the radio pulsars were known to a few arc seconds, time-resolved optical photometry was attempted and careful searches of the Sky Survey plates were made for visible evidence of the objects. With the exception of a few false alarms, the optical searches were negative until Cocke, Disney, and Taylor (1969) at the Steward

Observatory in Arizona found a pulsating stellar source in the center of the Crab Nebula, with a period equal to that of the radio pulsar. It was soon shown that the star emitting the optical pulses was that identified by Baade and Minkowski in 1942 as a remnant of the supernova explosion. Within the year, pulsations of this object had been detected at X-ray frequencies, and later at γ-ray energies. The most important recent development is the discovery of the first known pulsar in a binary system (Hulse and Taylor, 1975a). Accurate measurements of its orbit may help to distinguish between different gravitation theories, and its evolutionary history may shed light on the relation between pulsars and binary X-ray sources.

Pulsar Models

The bewildering and steady influx of observed facts during 1968 and 1969 gave pulsar theorists the opportunity to try out an equally bewildering array of proposed explanations. At first, the major effort was directed toward understanding the clock mechanism, the single most distinguishing characteristic of pulsars. Early reports had made it clear that the basic periodicities of most pulsars were stable to a precision better than one part in 10^7 over intervals of a few months. This fact by itself is sufficient to limit severely the range of tenable models, for it implies that objects with great inertia must be involved in the time-keeping process. The short periods showed that these objects must also be very compact compared to normal stellar objects. White dwarfs or the theoretically predicted neutron stars were obvious candidates.

Three distinct mechanisms were seen as possibilities for producing the periodic signals: radial pulsations (analogous to those in classical Cepheids), orbital motion, and rotation. Pulsations were the first to receive widespread attention, even though the observed periodicities seemed to be too fast for white dwarfs and much too slow for neutron stars. With the discovery of the Vela and Crab pulsars, both of which have periods well under 0.1 s, radial pulsations of white dwarfs were virtually ruled out. Theoretical models indicated that they could pulsate no faster than about once per second. Also, since a star's pulsation period is approximately proportional to $\rho^{-1/2}$, where ρ is the mean stellar density, the observed range of two orders of magnitude in period implied a density range too large for a single class of objects.

Orbiting white dwarfs or neutron stars were also soon ruled out—the white dwarfs because even a pair in contact would have an orbital period not less than about 1.7 s (Ostriker, 1968), and the neutron stars because the loss of energy in the form of gravitational radiation would result in a secular decrease of period, quickly leading to decay of the orbit. Observations of the Crab pulsar (Richards and Comella, 1969), the Vela pulsar (Radhakrishnan *et al.*, 1969), and longer-period pulsars (Cole, 1969; Davies *et al.*, 1969) had shown that in all cases the period was gradually increasing rather than decreasing. The problem of gravitational radiation is avoided if one postulates a small solid object with mass $\ll 10^{-6}\ M_\odot$ orbiting around a neutron star; however, tidal forces would destroy any such satellite.

Thus, only rotation remained as a viable explanation for the pulsar clock mechanism. Rotating white dwarfs are expected to be stable for rotation periods of one second or more (Ostriker, 1968); with shorter periods they would be destroyed by centrifugal forces. Furthermore, white dwarfs are easily visible at the estimated distance of the nearest pulsars, but, as mentioned above, no optical counterparts were found for these sources. In rotational models the stellar radius must be such that equatorial velocities do not exceed the velocity of light. For the Crab pulsar this implies a radius of less than about 1,700 km; neutron stars are the only known stellar configuration with a radius less than this value. Thus, the rapidly spinning neutron star, first considered by Pacini (1967) and first proposed as a pulsar model by Gold (1968), gained support as the simplest, most flexible, and least-tarnished method of obtaining periods in the observed range. Long before the discovery of pulsars it had been suggested by Baade and Zwicky (1934) that neutron stars would be formed in supernova explosions. The association of the two pulsars having the shortest periods known at the time with known supernova remnants gave further support to this interpretation.

If the radiation we observe as pulses is confined to a single narrow beam, then the pulse period is equal to the rotation period of the star. On this basis Gold correctly predicted that the period of pulsars should gradually increase. Further supporting evidence for the rotating neutron star model was provided by the fact that the rate at which the Crab pulsar loses rotational kinetic energy is equal (within the estimation errors) to the energy requirements of the Crab Nebula. Also the observed pulse duty cycles and the time-reversal symmetry of many pulse shapes can readily be accommodated within the basic "lighthouse beacon" model, whereas they are more difficult to obtain in pulsation or other types of models.

By the middle of 1969 there was general agreement on the superiority of the rotating neutron star model for explaining the regularity of pulsation. However, it has proved to be many times more difficult to understand the mechanism responsible for the pulses themselves. The observed radio flux densities, together with reasonable estimates of distance and upper limits to the sizes of emitting regions, in some cases imply apparent brightness temperatures higher than 10^{30} K. For incoherent emission there is the fundamental thermodynamic limitation that the brightness temperature $T_b \leq \varepsilon/k$, where ε is the particle energy and k is Boltzmann's constant; thus for $T_b \approx 10^{30}$ K, particle energies of 10^{26} eV or more would be required, many orders of magnitude larger than can be produced by any known mechanism. It is almost certain, therefore, that in order to produce the observed radio pulse intensities the emission mechanism must be highly coherent, with many particles radiating in phase. For all pulsars the rate of rotational kinetic energy loss (derived from the observed slowing-down rates) is more than sufficient to account for the observed radio pulses, and, in the case of the Crab and Vela pulsars, for the high-frequency pulsed radiation as well. However, as yet no mechanism for converting this energy into the pulses we observe has been generally accepted. Proposed emission mechanisms are discussed in Chapter 10.

Basic Observational Properties

At the present time 149 pulsars are known. All of these share a number of basic characteristics, the most important being the emission of broadband radio noise in the form of a periodic sequence of *pulses* (Figure 1-3). Observed pulse intensities vary over a wide range; sometimes entire pulses are missing. Nevertheless, regardless of the pulse intensity, the basic timing of pulses is periodic. If the instrumental time constant is reduced to about 1 ms, a more complex pulse structure is revealed.

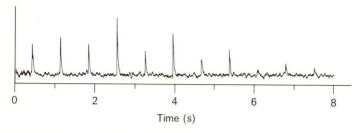

Time (s)

1-3 Chart record of individual pulses from one of the first pulsars discovered, PSR 0329 + 54. They were recorded at a frequency of 410 MHz and with an instrumental time constant of 20 ms. The pulses occur at regular intervals of about 0.714 s.

Individual pulses are often found to consist of two or more *subpulses*, which frequently overlap and have characteristic widths of one or two percent of the period. A further decrease in time constant to about 10 μs reveals that within the subpulses some pulsars show *microstructure*, which has a characteristic width of about 0.1 percent of the period. Because the basic timing mechanism of pulsars is identified with rotation of neutron stars, the time within a period is often described in terms of *longitude* (ϕ), so that the pulse period equals 360° of longitude. In terms of longitude, typical widths for subpulses are about 5°, and for micropulses about 0.3°.

In addition to varying in amplitude, subpulses vary in longitude or phase within the *pulse window* or *integrated profile*, the mean pulse profile obtained by adding many pulses synchronously with the pulse period. Despite the variable nature of subpulses in a given pulsar, the shape of the integrated profile is quite stable. Integrated profiles differ greatly from one pulsar to another. About half of the known pulsars have profiles with a single peak or *component*. Others have several components, sometimes partially overlapping. Two-component or "double" profiles are relatively common. Up to five components are seen in the profiles of some pulsars; in some such cases the components are symmetrically disposed about the profile center, so the profile still has a "double" appearance. The *equivalent width* of integrated profiles, defined as pulse energy divided by peak flux density, is generally about 10° of longitude (or three percent of the period), with the components of double profiles sometimes separated by as much as 40°.

Except for the pulsating X-ray sources in binary systems (Chapter 5), all of the known pulsars have been discovered at radio frequencies; apart from the possible detection of some longer-period pulsars at γ-ray frequencies, only the Crab and Vela pulsars (PSR 0531+21 and PSR 0833−45, respectively) are known to radiate outside radio frequencies. In general, the radio-frequency spectrum is quite steep; the spectral index, α, defined by

$$S = S_0 v^\alpha, \tag{1-1}$$

where S is flux density and v is frequency, is typically about −1.5. For most pulsars the spectrum becomes steeper at higher frequencies (above about 1 GHz); in some cases a spectral maximum is observed between 100 and 500 MHz.

As may be seen in Figure 1-4, observed pulsar periods lie between 0.03 and 4.0s, with a median value of about 0.65 s. There is an apparently significant dip in the distribution, centered on a period close to 1 s. This has suggested to some authors that pulsars are of two classes:

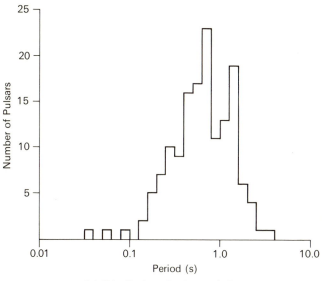

1-4 Distribution of pulsar periods.

those with short and those with long periods. The pulsar with the shortest known period (0.033 s) is of course the Crab pulsar; the pulsar with the second-shortest period (0.059 s) is a member of a binary system that has an orbital period of 7^h45^m. In all cases where accurate observations have been made, pulsar periods are found to be increasing in a regular way. Observed rates of change are typically about 10^{-15} s s^{-1}, or a few tens of nanoseconds per year. The regularly increasing periods imply that pulsars are unlikely to be older than the *characteristic time*, $T = P\dot{P}^{-1}$, typically about 10^7 years. For pulsars with shorter periods, rates of change are often larger and the characteristic times less; for the Crab pulsar, $T = 2,480$ years.

A plot of pulsar positions (in galactic coordinates) is shown in Figure 1-5. Despite the fact that several of the more more sensitive surveys have concentrated on latitudes within a few degrees of the galactic plane, the observed concentration of pulsars along the plane is a real effect and shows that the observed pulsars are located within our galaxy. The concentration of pulsars in the longitude quadrants toward the galactic center is also real, implying that, at least in the vicinity of the sun, the density of pulsars increases with decreasing galactocentric radius. The galactic distribution of pulsars will be discussed in more detail in Chapter 8.

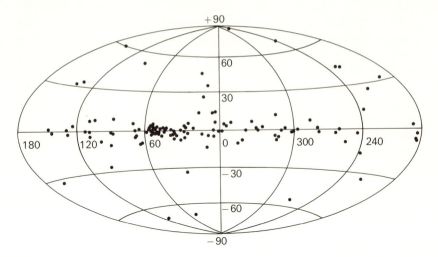

1-5 Distribution of pulsars in galactic coordinates, plotted on a Hammer equal-area projection.

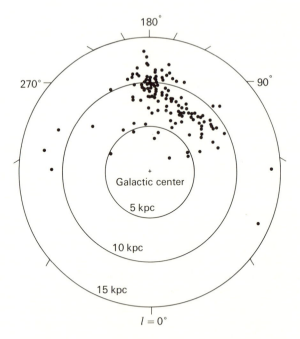

1-6 Distribution of pulsars projected onto the galactic plane. In most cases distances are derived from dispersion measures. Twenty-one pulsars within 2 kpc of the sun have been omitted for clarity.

As with most astronomical objects, accurate distances for pulsars are difficult to obtain. For most pulsars, distances can be estimated only from the dispersion suffered by the pulse during its propagation through the interstellar medium. The *dispersion measure* is proportional to the column-density of electrons in the line of sight to the pulsar, so if the electron-density distribution is known, distances can be computed. In practice, independently measured pulsar distances are used to determine the mean interstellar electron density and so to calibrate the distance scale for other pulsars. Distances obtained in this way range between about 100 and 20,000 pc* with the more distant sources always being within a few degrees of the plane. Because of uncertainty in the electron-density distribution, the distance for an individual pulsar may be considerably in error, perhaps by as much as a factor of two. However, on a statistical basis, they appear to be reasonably accurate. For most pulsars the computed z distance (the perpendicular distance from the galactic plane) is less than 300 pc, so the overall galactic distribution is very much disk-shaped. The distribution of pulsars projected onto the galactic plane is shown in Figure 1-6. No spiral structure is evident, no doubt mostly as a result of errors in the assigned distances.

A list of the principal characteristics of the 149 pulsars known at the time of writing is given in the appendix.

* pc: parsec; 1 parsec = 3.26 light-years = 3.086×10^{13} km.

2

Characteristics of Integrated Pulse Profiles

Soon after the discovery of pulsars it was recognized that the mean or integrated pulse profile for any given pulsar was very stable and had a characteristic shape. Early observations also showed that, in many cases, the profiles were highly polarized, with 100 percent linear polarization being observed for some pulsars. In this chapter we shall describe the properties of integrated profiles, including their shape, intensity, and polarization, and the time- and frequency-dependence of these properties.

Shape

One of the most important properties of a pulsar is the shape of its integrated profile. Because of the improvement in signal-to-noise ratio inherent in the averaging process, good quality integrated profiles, often at several different frequencies, have been obtained for most of the known pulsars. These observations show that profiles are often rather complex, with several components or identifiable peaks, and that each pulsar has a unique profile. Integrated profiles for 45 pulsars, showing the wide variety of observed shapes, are given in Figure 2-1. Most profiles

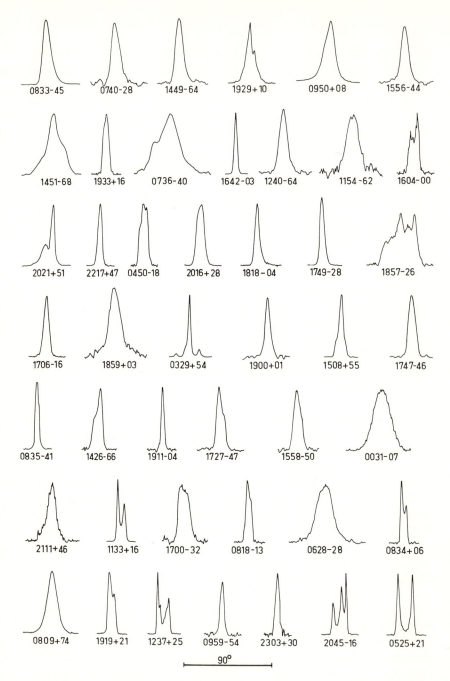

2-1 Integrated pulse profiles for 45 pulsars, all plotted on the same longitude scale (a 90°
bar is given in the bottom of the figure). These profiles were recorded at frequencies
between 400 and 650 MHz, and are arranged in order of increasing pulse period.

are dominated by a single component, but multiple-component profiles are also common. The basic shape of multiple-component profiles is usually "double," that is, two main peaks having steep outer edges separated by a saddle region. For example, PSR 1237+25 (Figure 2-1, bottom row) has five distinct components, but is basically double in form. In this pulsar the location of the components is nearly symmetrical about the profile center, but in others, e.g., PSR 2045−16 (Figure 2-1, bottom row), the central components are offset.

The shape of integrated profiles is generally somewhat frequency-dependent; however, the basic character of a pulsar's profile usually remains the same at all frequencies. Therefore pulsars may reasonably be divided into two broad groupings, according to whether they have "single" or "double" profiles, called respectively Type S (simple) and Type C (complex) by Taylor and Huguenin (1971). Since more observations are available at frequencies around 400 MHz than at other frequencies, profile shapes obtained at this frequency are conventionally used for purposes of classification. This division according to profile shape correlates with other properties. For example, Type C pulsars tend to have long periods, in most cases greater than one second (Figure 2-1); they seldom have very small values of the parameter $P\dot{P}$, which is related to the magnetic field strength at the surface of the neutron star (Chapter 9); they usually have strong linear polarization; and they exhibit a smooth variation in the polarization position angle across the profile. Type S pulsars, on the other hand, tend to have short periods, and often have low values of $P\dot{P}$, weak polarization, and discontinuous changes in the position angle across the profile.

Another characteristic feature of pulsars, which overlaps the above categories, is the phenomenon of *drifting subpulses* (p. 40). Pulsars with drifting subpulses are called Type D or, when more detail is required, Type SD or CD, indicating a simple or complex mean profile. The most highly organized drifting patterns occur in Type SD pulsars.

It is clear from Figure 2-1 that many pulsars have two or more unresolved or overlapping components. Backer (1976) has proposed a classification scheme in which pulsars with two overlapping components are grouped in a "double-unresolved" category. He also suggests that pulsars with three or more components should be grouped separately as Type M (multiple).

Although the pulsed energy from most pulsars is confined to a small fraction of the period, there are a number of exceptions. In several pulsars an additional pulse component, the *interpulse*, is situated approximately half-way between the main pulses. For example, the Crab pulsar, PSR 0531+21, has three components: the main pulse and its precursor

TABLE 2-1

Pulsars with interpulses at radio frequencies

PSR	Period (s)	Interpulse energy (percent of main pulse)	Separation of main pulse and interpulse (deg)
0531 + 21	0.033	36%*	145
0823 + 26	0.531	0.5	180
0904 + 77	1.579	20	180
0950 + 08	0.253	1.8	150
1055 − 52	0.197	85	155
1929 + 10	0.226	2	174

* For PSR 0531 + 21, the main pulse energy includes that of the precursor.

component and an interpulse, which has an energy comparable to that of the main pulse and is situated about 13 ms or 145° of longitude later. In half the pulsars known to have interpulses the interpulse energy is only a few percent of the main pulse energy (Table 2-1). It is possible that the true period of PSR 0904 + 77 is half the adopted value, but this is very unlikely for any of the others.

The pulsed emission of most pulsars is confined to a rather narrow longitude range, but exceptions to this rule are observed. For example, significant emission is seen in the shorter of the two intervals between the main pulse and interpulse of PSR 0950 + 08; this emission reaches a minimum value of about 0.1 percent of the main pulse intensity (Figure 2-2). There may also be a nonvarying component to the emission from pulsars. In normal synchronous averaging such a component would be removed by the baseline-fitting procedure. A search for such emission by Huguenin and co-workers (1971) using interferometric techniques yielded upper limits comparable to the mean flux density (pulse energy divided by the period) for several pulsars. In some pulsars the integrated profile is much broader than usual. For example, in PSR 1541 + 09 the pulse width at half-power is 33° and emission extends over more than half the period (Figure 2-2). With a width at half-power of 75°, PSR 1911 + 03 has the broadest known profile.

These wide profiles are intrinsic to the pulsar and do not result from propagation effects in the interstellar medium. At low frequencies pulse profiles are often affected by scattering of the radiation by irregularities in interstellar electron density. Propagation delays result in a smearing of the pulse energy into an exponentially decaying pulse tail. If the scattering is severe, delays can exceed the pulse period, thereby causing a nonvarying flux component and a decrease in pulsed energy.

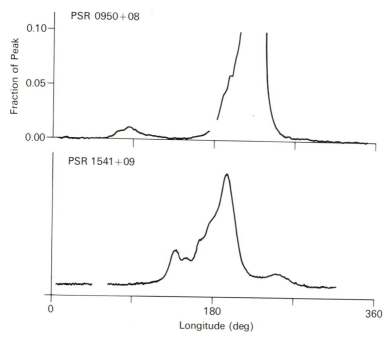

2-2 Truncated integrated profile for PSR 0950 + 08 and integrated profile for PSR 1541 + 09 recorded at a frequency of 430 MHz. These pulsars are exceptional in having wide pulse profiles. [After Backer *et al.*, 1973.]

Interstellar dispersion also produces smearing of the pulse profile, often limiting the time resolution, especially for low frequency observations. This effect can be largely overcome by splitting the signal into a number of narrow frequency bands and then recombining with appropriate delays. Two different techniques have been employed. The first uses postdetection delays (Taylor and Huguenin, 1971), the second predetection sampling followed by computer analysis (Hankins, 1971). Using such procedures, Hankins (1973) detected a narrow notch of width about 0.4 degrees of longitude near the peak of the profile of PSR 1919 + 21; a similar but weaker notch was found about 2.2 degrees after the peak. Further analysis by Cordes (1975a) showed that the fractional depth of the main notch is a strong function of frequency, being about 0.5 at 74 MHz and falling to zero at about 250 MHz.

The equivalent width (pulse energy divided by peak flux density) is commonly used to specify the longitude extent of integrated profiles. This quantity is plotted against period in Figure 2-3 for most of the known pulsars. Although there is considerable scatter in the data, pulse

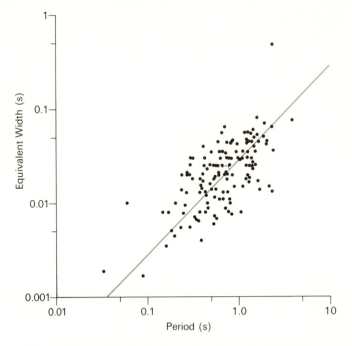

2-3 Equivalent widths (or in some cases half-power widths) of the
integrated pulse profiles of most known pulsars plotted against
pulse periods. The line represents a mean pulse width of 10° of
longitude.

widths are proportional to period; that is, when expressed in terms of
longitude, pulse widths are approximately constant. The mean equivalent
width is approximately 10° of longitude or 3 percent of the period. This
result is consistent with the interpretation of the integrated profile as a
cross section of a beam emanating from a rotating star.

In general, the shape of the integrated profile for a given pulsar is
slightly frequency dependent. For pulsars having two or more components the separation of these components is normally proportional to
some power of the frequency:

$$\Delta\phi \sim \nu^p. \qquad (2\text{-}1)$$

As shown in Figure 2-4, the separation index, p, is typically about -0.2
at low frequencies. In many cases there is a break frequency above which
the variation remains power-law but with a more positive index. Table
2-2 lists the component separation, break frequency, separation index,

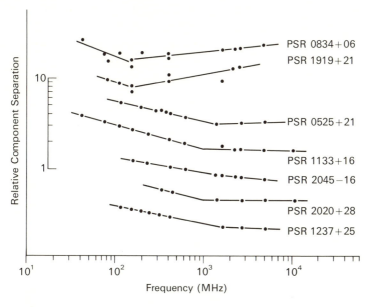

2-4 Separation between outer components of pulsars with "double" integrated profiles. [After Sieber *et al.*, 1975.]

and number of profile components for several pulsars. If a profile has more than two components, the proportional spacing is normally independent of frequency. An exception to this is PSR 0329+54, for which the separations between the first and third (main), and third and fourth (trailing) components have different degrees of frequency dependence. This fact, together with the polarization effects discussed below, suggests that in this pulsar there may be overlapping emission from independent regions.

In addition to decreases in component separation with increasing frequency, observations by Sieber, Reinecke, and Wielebinski (1975) show that, on the average, the individual components themselves decrease in width at higher frequencies. Observed half-power widths of apparently unblended (nonoverlapping) components at 4.9 GHz are typically 15 percent less than those at 2.7 GHz. In a few cases widths at 4.9 GHz are greater than those at the lower frequency; this may result from the undetected blending of two or more components.

Different components of Type C profiles normally have slightly different spectral indexes, α (Equation 1-1); this also leads to a variation with frequency in the shape of integrated profiles. There is no consistent location in the profile for the component with the steepest spectrum.

TABLE 2-2
Components of integrated profiles of selected pulsars

PSR	Number of components	Separation at 400 MHz[a] (deg)	Break frequency (MHz)	Separation index, p	
				Low freq.	High freq.
0301 + 19	2	11.1		−0.30	
0329 + 54 (1−3)	4(5)	13.3	1100	−0.18	−0.08
0329 + 54 (3−4)		9.8	900	−0.08	+0.04
0450 − 18	3	14			
0525 + 21	4	14.5	1390	−0.21	+0.06
0628 − 28	2	5[b]			
0736 − 40	3	38[c]			
0833 − 45	2	7.5[b]			
0834 + 06	2	4.5	135	−0.36	+0.11
0950 + 08	2	4		−0.5	
1133 + 16	2	6.6	970	−0.26	0.00
1237 + 25	5	11.4	1410	−0.18	−0.01
1451 − 68	3	25[c]			
1508 + 55	3	9		−0.25	
1919 + 21	2	4.8	145	−0.34	+0.18
1929 + 10	3	5.0		−0.28	
1933 + 16	2	5			
2016 + 28	2	4			
2020 + 28	3	10.8	960	−0.20	+0.01
2021 + 51	2	7			
2045 − 16	3	13.3	1500	−0.15	−0.07
2319 + 60	3	13.5[b]			

[a] Except for PSR 0329 + 54, separations are between outermost components.
[b] Measured at 1400 MHz.
[c] Measured at 630 MHz.

For components of a given pulsar the difference between indexes ($\Delta\alpha$) is typically in the range 0.1–0.4. For example, for PSR 2045 − 16 the spectral indexes of the second and third components relative to the first are −0.4 and +0.1, respectively. In some cases $\Delta\alpha$ is larger. At frequencies above about 500 MHz the spectral index of the precursor component of the Crab pulsar profile is less than −5.0, compared to a value of approximately −2.8 for the main pulse. Below this frequency the spectral indexes for these two components are more nearly equal.

In general, the number of identifiable components in the profile of a given pulsar does not change with frequency. However, in a few cases, at very high or very low frequencies, a component may become undetectable and/or apparently new components may appear. For example, at frequencies around 400 MHz PSR 1642−03 has a simple, almost Gaussian, profile; but observations by Sieber, Reinecke, and Wielebinski (1975) show that at frequencies above about 2 GHz it has a three-component profile similar to that of PSR 1508+55 (Figure 2-1, fourth row). Because of these variations, it is sometimes difficult to identify the same component at widely differing frequencies; this can lead to uncertainties in calculations of the dispersion measure.

Energies

Observed pulse energies vary on many different time scales. Variations observed on scales from a few minutes to several hours can result from the effects of interstellar scattering (Chapter 7), but those observed on shorter and longer scales are generally intrinsic. Here we shall consider the effects on longer time scales; pulse-to-pulse variations will be discussed in Chapter 3.

At frequencies around 400 MHz the stronger pulsars have an equivalent mean flux density of about 0.1 Jy.* As their duty cycle is about three percent, this corresponds to a peak flux density of a few Janskys. Pulse energies, when averaged over intervals of a few hundred pulse periods, are relatively stable over intervals of several hours (after removal of the effects of scattering) but may vary by a factor of two or more from day to day (McLean, 1973). On longer time scales the variations are even larger; Figure 2-5 shows the pulse energies of five pulsars recorded over a two-year period. Variations in intensity of up to an order of magnitude occur with characteristic fluctuation times of 20–70 days. Provided at least a few hundred pulses are added together, the shape of an integrated profile at a given frequency is independent of fluctuations in the mean pulse energy.

Variations in mean pulse energies make it difficult to determine the spectrum of the radio emission from pulsars. To obtain reliable results, simultaneous observations at different frequencies and/or averaging of many independent observations are required. Investigations of this type show that, although pulsar spectra have several different forms, at high radio frequencies the spectral index is always negative (Sieber, 1973; Backer and Fisher, 1974). Examples of observed spectra are shown in Figure 2-6. About half of the pulsars have apparently straight (power-law) spectra, at least within the frequency ranges observed. The spectra

---

* Jy: Jansky; 1 Jy = 10^{-26} W m^{-2} Hz^{-1}.

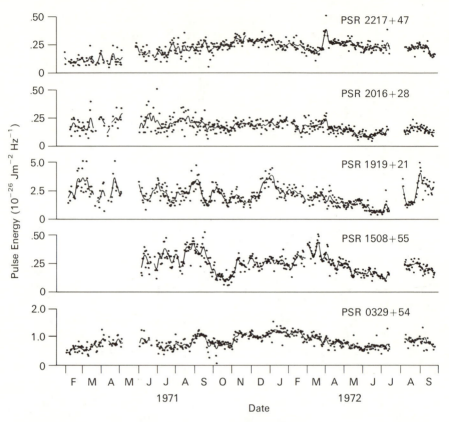

2-5 Pulse energies of five pulsars observed at a frequency of 156 MHz over a two-year period. Each point represents the pulse energy averaged over 2–4 hours of observation. [From Huguenin *et al.*, 1973.]

of the remainder are less steep at lower frequencies, with either two straight segments, as in PSR 1133 + 16, or a low frequency cutoff, as in PSR 0329 + 54. In some pulsars, for example PSR 1929 + 10, both effects are observed. This spectral break between two power-law segments appears to be related to the similar effect observed for component separations (Figure 2-4). For example, Sieber (1973) finds the following spectral-break frequencies: PSR 0525 + 21, 1,080 MHz; PSR 0834 + 06, 140 MHz; PSR 1133 + 16, 1,400 MHz; PSR 1919 + 21, 260 MHz. These values may be compared with the corresponding frequencies in Table 2-2.

The distribution of observed spectral indexes for a sample of the known pulsars is shown in Figure 2-7. Indexes are given for pulsars with straight spectra (including some with a low-frequency cutoff) and those that have a break in slope (two straight segments). The spectral indexes for most

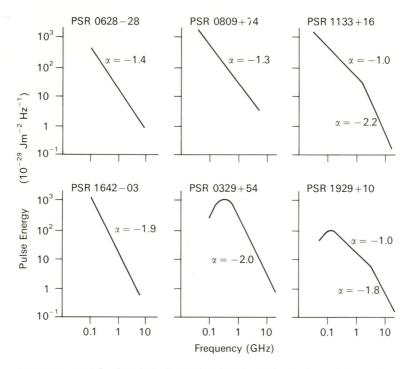

2-6 Pulse spectra for six pulsars, illustrating the different forms observed. Power-law spectral indexes (α) are given beside each curve. [After Sieber, 1973.]

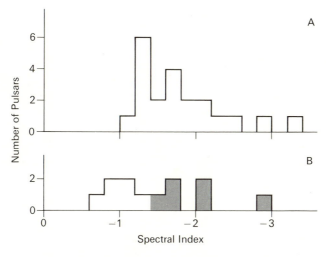

2-7 Histogram of spectral indexes. **A**. Pulsars with straight spectra, including some with a low-frequency cutoff. **B**. Six pulsars with spectra consisting of two straight segments (a break in slope). The first four bars represent the lower-frequency segment, the grey bars the higher-frequency segment. [After Sieber, 1973.]

pulsars are between -1 and -3, with the peak in the distribution at about -1.5. For spectra with a break in slope, indexes below the break are close to -1.

The observed low-frequency cutoffs are likely to be intrinsic to the emission mechanism, except when scattering is important (see below). They do not have the exponential form expected for free-free absorption in the interstellar medium; furthermore, there is no correlation between cutoff frequency and dispersion measure. Elitzur (1974) pointed out that plasma frequencies are reduced by relativistic streaming of particles in the pulsar magnetosphere (Chapter 9), so that the observed cutoffs are unlikely to result from plasma effects in this region.

Low-frequency cutoffs resulting from a propagation effect—interstellar scattering—are observed in several short-period pulsars. For example, rather sharp cutoffs are observed in the spectra of the Crab pulsar (PSR 0531+21) at about 100 MHz (Rankin *et al.*, 1970) and the Vela pulsar (PSR 0833−45) at about 300 MHz (Komesaroff *et al.*, 1972). In each case the pulsed energy is smeared to form a nonpulsed source, the spectrum of which is a continuation of that of the pulsar. Observed spectral cutoffs in most longer-period pulsars do not result from scattering, but for the more distant pulsars this effect is likely to be important at frequencies below 200–300 MHz. (Interstellar scattering is discussed further on p. 137ff.)

Polarization

One of the striking characteristics of pulsar emission is the high degree of polarization it frequently possesses. The integrated profiles of a number of pulsars have essentially complete linear polarization, implying both complete polarization of all individual pulses and stable polarization for all emission at a given longitude. Circular polarization is also observed, but in the integrated profile it rarely exceeds 20 percent of the total intensity. The polarization characteristics of the integrated profiles of four pulsars are shown in Figure 2-8. The Vela pulsar, PSR 0833−45, has almost total linear polarization, whereas PSR 1642−03 is very weakly polarized, especially at its trailing edge. Pulsars with drifting subpulses (see p. 40) also have weakly polarized integrated profiles. The degree of polarization for Type C pulsars is normally greatest at the inner edges of the outer components and least in the extreme wings of the profile. For PSR 1133+16, which has rather closely spaced components, this effect results in maximum polarization (about 60 percent) in the saddle region between the components. For pulsars of other types the region of maximum polarization may occur anywhere within the profile. The polarization characteristics of the integrated profiles

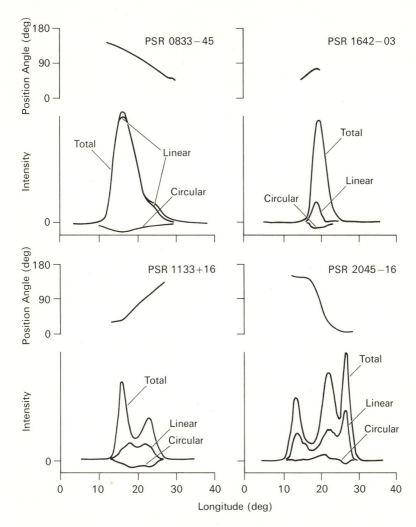

2-8 Polarization characteristics of the integrated profiles of four pulsars. In each graph the two curves under the total intensity profile represent the linear and circular polarization intensities; position angles are plotted in the upper part of each graph. The profile for PSR 0833 − 45 was recorded at a frequency of 1665 MHz; the profiles of the other three were recorded at 410 MHz. [After Manchester, 1971a.]

of a number of pulsars, measured at a frequency close to 400 MHz (in most cases), are given in Table 2-3. These data illustrate the wide range of fractional polarization observed.

In contrast to most radio sources, the fractional polarization of pulsar integrated profiles normally decreases with increasing frequency. Three

TABLE 2-3
Polarization characteristics of the integrated profiles of selected pulsars

PSR	Type	Linear polarization at 400 MHz	Extent of position angle variation (deg)	Maximum position angle gradient $\lvert d\psi/d\phi \rvert_{max}$
0031−07	SD	3%	—	—
0301+19	CD	50	120	13
0329+54	(S)[a]	16	(180)[b]	28
0525+21	C	45	150	42
0531+21	(S)	25	∼20	∼1
0628−28	S	48[c]	90	4.5
0736−40	(S)	12[d]	(110)	—
0809+74	SD	6	(90)	1.8
0833−45	S	95[c]	85	6.0
0834+06	C	10	(65)	19
0950+08	S	20	35	1.8
1133+16	C	33	105	10
1154−62	S	20[d]	40	2
1237+25	C	50	180	60
1240−64	S	11[d]	60	4.1
1451−68	(S)	12[d]	(145)	20
1508+55	(S)	20	180	19
1556−44	S	50[d]	65	9
1642−03	S	13	20	9
1700−32	S	30[d]	135	40
1919+21	CD	10	25	12
1929+10	S	78	35	1.6
1933+16	S	20	35	6.5
2016+28	SD	12	(120)	10
2020+28	C	40	(120)	10
2021+51	S	55	50	3.5
2045−16	C	40	145	37
2319+60	C	40[c]	75	10

[a] Parentheses indicate that the pulsar is not clearly identified as any one type, but appears to be closest to the type listed.
[b] Parentheses indicate that the position angle for the pulsar changes discontinuously (see Figure 2-10).
[c] Measured at 1400 MHz.
[d] Measured at 630 MHz.

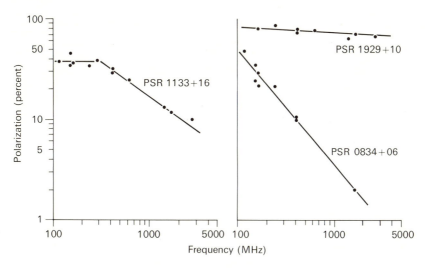

2-9 The three types of frequency dependence of fractional polarization of integrated profiles. [After Manchester, Taylor, and Huguenin, 1973.]

forms of frequency dependence of polarization are observed (Figure 2-9). In several pulsars, e.g., PSR 1133+16, the fractional polarization is constant up to some critical frequency, and above this frequency varies as v^{β} with $\beta \approx -1$. The critical frequencies do not appear to be related to the break frequencies listed in Table 2-2. For other pulsars there is no critical frequency within the observed range. In some cases, e.g., PSR 1929+10, the fractional polarization is almost constant; in these cases the critical frequency may be higher than the maximum frequency observed. In other cases, e.g., PSR 0834+06, the fractional polarization varies approximately as v^{-1} throughout the whole of the observed range. Observed values of β range between -0.4 and -1.2 in different pulsars (Manchester, Taylor, and Huguenin, 1973).

It is clear from Figure 2-8 that the position angle of the linearly polarized emission (the orientation of the electric vector on the sky) varies throughout the profile. In most pulsars this variation is continuous, and is often of the approximately linear form shown in Figure 2-8 for PSR 0833−45 and PSR 1133+16. The position angle curve for Type C pulsars usually has a characteristic S-shape, with rapid variation near the center of the profile and slow variation in the wings; the profile for PSR 2045−16 in Figure 2-8 is an example of this type of curve. For all pulsars the total change of position angle throughout the profile is less than or about 180 degrees; values of this parameter together with the maximum gradient of the position angle curve are given in Table 2-3 for the pulsars listed.

For Type C pulsars there is a correlation between these quantities: the larger the maximum gradient, the closer the total variation is to 180°. As will be discussed in Chapter 10, this type of position angle variation is exactly that expected for the projection of a vector fixed with respect to a rotating system. It is notable that these S-shaped curves are always symmetrically located within the integrated profile, with their center of symmetry either exactly or very close to midway between the outer edges of the profile.

The variation in position angle of a given pulsar is the same at all frequencies. As discussed earlier, component separations normally increase with decreasing frequency (see Figure 2-4). However, the position angle curve does not expand at lower frequencies; the gradient at a given longitude is independent of frequency. Consequently, the total change of position angle throughout the pulse is normally somewhat larger at lower frequencies. These results, together with the symmetrical form of the position angle curves, strongly suggest that the observed position angle variations are determined by the emission mechanism and the line-of-sight geometry, rather than by propagation effects occurring away from the source region.

Such variations in position angle and fractional polarization throughout the profile as are shown in Figure 2-8 cannot result from the overlapping of components polarized at different position angles. In the case of overlapping components one expects a drop in fractional polarization in the region of overlap; this is not observed in these profiles. However, in some pulsars, e.g., PSR 0329 + 54 and PSR 2020 + 28, rapid position angle changes accompanied by a drop in fractional polarization are observed (Figure 2-10). Essentially discontinuous changes of about 90° in position angle are observed at two longitudes in both PSR 0329 + 54 and PSR 2020 + 28. In the profile of each pulsar one of the discontinuities is between partially overlapping components of different position angle, but the other occurs within an apparently single component. These results suggest that there is overlapping of radiation beams either from two separate emission regions or from two emission modes generated in one region, with the position angle of the linearly polarized emission in the two beams being close to orthogonal. This effect is closely related to the orthogonal polarization observed in individual pulses from many pulsars (see p. 51ff.)

As mentioned above, very high degrees of circular polarization are not observed in integrated profiles. For the pulsars shown in Figure 2-8 the maximum fractional circular polarization (near the center of the profile of PSR 1133 + 16) is about 20 percent. In most cases one sense of circular polarization (either right or left) predominates throughout a given profile. However, several pulsars exhibit one or more reversals of sense throughout the profile; two of these are shown in Figure 2-11.

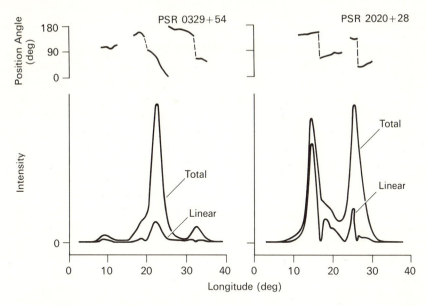

2-10 Linear polarization characteristics of the integrated profiles of PSR 0329 + 54 and PSR 2020 + 28 (recorded at frequency 410 MHz). As in Figure 2-8, the linear polarization intensity is plotted in the curve under the total intensity profile. The abrupt changes in position angle apparently result from overlap of independent components. [From Manchester, 1975.]

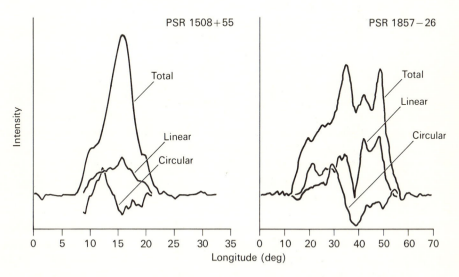

2-11 Integrated profiles of two pulsars showing reversals in the sense of circular polarization throughout the profile. Polarized intensities are plotted as in Figure 2-8. [After Manchester, 1971a; and Hamilton *et al.*, 1977.]

Stability

In general, integrated profiles remain stable in shape and polarization on long time scales; this is of course the reason why they are an important feature of pulsar emission. Individual pulses vary greatly in shape, intensity, and longitude from one pulse to the next (Chapter 3). In view of this, the stability of an integrated profile clearly depends on the number of pulses included in it. As a measure of profile stability, Helfand, Manchester, and Taylor (1975) have computed the average cross-correlation coefficient between integrated profiles of a given pulsar and a standard profile for that pulsar (obtained by adding all the available data) as a function of the number of pulses included in the integrated profiles. For pulsars of Type S and Type D, correlation coefficients (ρ) generally exceed 0.9995 when data are summed for 500 pulse periods, whereas Type C pulsars reach a stable form more slowly. For example, a data span of about 2,500 periods is required for profiles of PSR 1133 + 16 to reach a correlation coefficient of 0.9995. For most pulsars the quantity $1 - \rho$ has an approximate power-law dependence on the number of pulses included in the profile. Type D pulsars have the strongest dependence of $1 - \rho$ on the number of pulses averaged, with an average slope (on a logarithmic plot) of about -1.3; that is, they reach a stable form very quickly. The slope is least for Type C pulsars, especially those with more than two components. For example, the line for PSR 1237 + 25 has a slope of -0.55, close to the value of -0.5 expected for purely random subpulse variations. Type S pulsars are intermediate with slopes of about -0.8. The average correlation between individual pulses and the standard profile is also dependent on the profile shape, ranging from 0.94 for PSR 1929 + 10 to 0.55 for PSR 1237 + 25.

Part of the reason for the low average correlation coefficient in PSR 1237 + 25 is that the integrated profile occasionally changes to a second stable form. This phenomenon, known as *mode changing*, was first observed by Backer (1970a). At irregular intervals, typically separated by one to two hours, the pulsar abruptly changes to a second mode in which the third component is strong and the fourth and fifth components are almost absent. The pulsar normally remains in this mode for a few tens or hundreds of periods before abruptly reverting to its normal shape. Short segments of the second mode appear to be more common than longer ones.

Similar changes are observed in PSR 0329 + 54 and PSR 1604 − 00. Several possible modes can occur in PSR 0329 + 54. In one of these, observed at a frequency of 2.7 GHz by Hesse (1973), component 2 is enhanced, component 4 is weak or absent, and a new component (component 5) appears in the saddle region between components 3 and 4

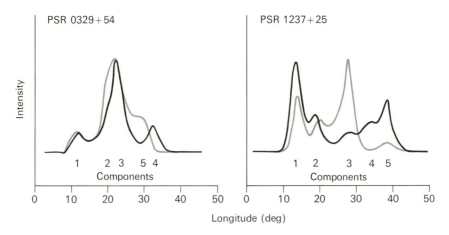

2-12 Integrated profiles of two pulsars illustrating mode changing. The normal
mode is represented by the black line, the abnormal by the gray line. The
data for PSR 0329 + 54 were recorded at a frequency of 2695 MHz (Hesse, 1973)
and for PSR 1237 + 25 at 285 MHz.

(Figure 2-12). In observations at a frequency of 10.7 GHz, Hesse and
co-workers (1973) found another mode, in which components 2 and 4 are
enhanced; component 4 is stronger by more than a factor of ten and
dominates the profile. At frequencies around 400 MHz component 4 is
normally twice as strong as component 1 (see Figure 2-10). Lyne (1971)
found yet another mode in which the relative strength of these two components was reversed with component 1 twice as strong as component 4.
Clearly there is a complex relationship between the various components
in this pulsar.

No long-term or secular change in the shape of integrated profiles or
their intrinsic polarization has ever been detected. Helfand, Manchester,
and Taylor investigated the long-term stability of the profiles of four
pulsars over an interval of about three years and found no case in which
the variations exceeded those expected on the basis of the number of
pulses included in the profiles. For the mode-changing pulsars, no
significant change in the shape of the normal-mode profile has been
observed. Timing observations (see Chapter 6) show that the longitude
(or phase) stability of the integrated profile is also very high, with observed
fluctuations in profile phase often less than 10^{-4} periods. In fact, the
observed fluctuations appear to result from variations in the rotation rate
of the neutron star, so 10^{-4} periods is an upper limit to fluctuations of
profile phase with respect to a fixed longitude on the rotating neutron
star.

3

Characteristics of
Individual Pulses

Individual pulses from a given pulsar vary greatly in intensity, shape, and polarization from one period to the next. In general, they do not have the same form as the integrated profile. Variations are often random in character, but periodic changes are also observed, particularly in the pulse intensity. In this chapter we shall discuss the characteristics of individual pulses, including subpulses, micropulses, pulse-to-pulse intensity fluctuations, and drifting subpulses.

Subpulses

Individual pulses normally consist of one or more subpulses. These subpulses, which appear to be basic units of emission, typically have a rather simple, almost Gaussian shape and a width between 3 and 10 degrees of longitude. Subpulses occur at various longitudes within the integrated profile, and often overlap when two or more subpulses are present in an individual pulse. Components or peaks are formed in the integrated profile when subpulses are stronger and/or occur more frequently at a given longitude. Longitude–time diagrams of the intensity

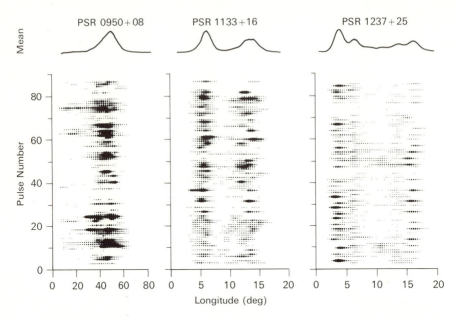

3-1 Longitude–time diagrams for three pulsars showing the variations in shape and intensity of a series of individual pulses. Each horizontal series of dots represents one pulse; the size of the dots indicates intensity. Successive pulses are plotted upwards on the diagram; integrated profiles are shown at the top. Subpulses, or bursts of enhanced emission covering 3–10 degrees of longitude, can be seen in most pulses. [After Taylor *et al.*, 1975.]

variations in sequences of individual pulses from three pulsars are shown in Figure 3-1. These diagrams show that subpulses are generally narrow compared to the integrated profile, and that they occur preferentially at certain longitudes. This is especially true for multiple-component pulsars, such as PSR 1133 + 16 and PSR 1237 + 25.

Mean subpulse widths are plotted against widths of the integrated profiles of a number of pulsars in Figure 3-2. Subpulse widths are comparable to the integrated-profile widths only for those pulsars whose integrated profiles are dominated by a single strong component, e.g., PSR 0329 + 54 and PSR 1642 − 03, and are usually less for those pulsars with periods greater than 0.75 s. A plot of subpulse widths versus periods shows that the widths are approximately proportional to $P^{1/2}$, although there is a large scatter. This contrasts with the equivalent widths of integrated profiles, which are proportional to period (see Figure 2-3, p. 18, and suggests that the subpulse profile may represent a time variation in intensity rather than a beam profile. Indeed, Figure 3-2 could be interpreted as showing that subpulse widths represent a time scale for emission that is essentially independent of period. Observed subpulse widths are

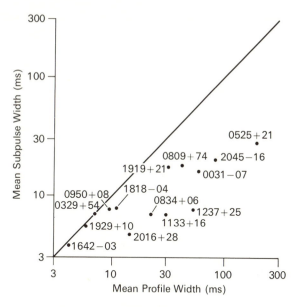

3-2 Mean half-power width of subpulses plotted against the
half-power width of the integrated profile for 14 pulsars.
Near the diagonal line, all subpulses have widths similar
to that of the integrated profile. [After Taylor *et al.*, 1975.]

of course restricted by the beaming process that produces the integrated profile. This cutoff (represented by the diagonal line in Figure 3-2) is especially significant for the shorter-period pulsars.

Subpulse widths are not strongly dependent on either frequency or pulse longitude within the integrated profile, although subpulses occurring between components of multiple-component profiles tend to be wider than average. There is also some correlation between subpulse intensity and width, with the stronger subpulses tending to be narrower. Observations at widely spaced frequencies show that subpulse intensities are very well correlated over wide frequency intervals. The subpulse emission process is therefore broadband, with the bandwidths typically exceeding 200 MHz. Robinson and co-workers (1968) showed that the spectra of most individual pulses from PSR 1919+21 were similar and hence similar to that of the integrated profile. It is likely that this is true for most pulsars. For pulsars with multiple-component profiles this high correlation of intensities at different frequencies, together with the observed frequency dependence of component separation (see Figure 2-4, p. 19), implies that the longitude of subpulses varies with frequency in the same way as the component longitude. That is, the longitude interval between a given subpulse and the profile center is greater at low frequencies than at high frequencies, with an approximately $v^{-0.25}$ dependence.

Intensity Fluctuations

It is clear from Figure 3-1 that subpulse intensities vary greatly from one pulse to the next. Here we shall describe two characteristic types of fluctuation, pulse nulling and periodic intensity variations; drifting sub-pulses are described in the next section.

Pulse *nulling* is a relatively common phenomenon in which the pulse intensity suddenly drops to a low value for a few pulses and then abruptly returns to normal. In multiple-component profiles all components drop in intensity. Ritchings (1976) finds that the intensity of null pulses is less than one percent of the normal pulse intensity. Short nulls of one or two missing pulses occur frequently in many pulsars but are especially pro-minent in PSR 0834 + 06 and PSR 1929 + 10. For example, in a sequence of 5,000 pulses from PSR 1929 + 10, Backer (1970b) found about 50 nulls, in each of which one or two pulses were missing. Longer nulls of 3–10 pulse periods are also common—several examples of such nulls can be seen in Figure 3-1. PSR 0031 − 07, a pulsar with highly organized drifting subpulses, is typically in a null state about 50 percent of the time, with pulse bursts of 10–100 pulses separated by nulls of similar duration. Another pulsar that shows drifting subpulses, PSR 1944 + 17, is in a null state for more than 75 percent of the time. The fraction of time that a pulsar is in a null state has been shown by Ritchings (1976) to be related to the pulsar period and period derivative; this will be discussed further in Chapter 10 (p.233).

The occurrence of these longer nulls appears to be random—no significant periodicities are observed. However, power-spectral analyses of sequences of pulse energies show that periodic fluctuations in pulse intensity do exist in a number of pulsars. In some pulsars these fluctua-tions appear to be related to the nulling phenomenon, whereas in others they are related to drifting subpulses. Fluctuation spectra show that narrow line features, representing strongly periodic fluctuations, are fairly common, particularly for the longer-period Type C and Type D pulsars (Figure 3-3). Out of the seven pulsars in Figure 3-3 with periods less than 0.75 s, only one has a strong periodic modulation, whereas five out of seven with periods greater than 0.75 s have strong features. For PSR 0834 + 06 and PSR 0943 + 10 the line features are narrow ($Q \gtrsim 50$) and are at frequencies close but not equal to the Nyquist frequency of 0.5 cycles/period, that is, an approximately alternate-pulse modulation. For PSR 0809 + 74 the strong feature at 0.090 cycles/period is clearly

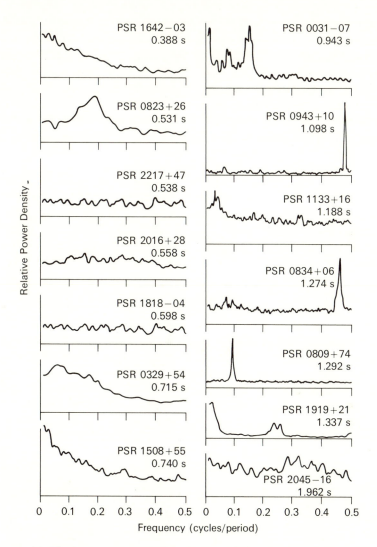

3-3 Fluctuation spectra for 14 pulsars, arranged in order of increasing
period, from 0.388 to 1.962 s. Zero levels of the spectra are given by
horizontal lines. [After Taylor and Huguenin, 1971.]

related to the drifting subpulses described in the next section. The shorter-
period pulsars are characterized by featureless spectra that are either
flat (representing random fluctuations) or rising toward low frequencies
(representing relatively slow aperiodic modulation).

The fluctuation spectrum for PSR 1237+25 has a relatively strong
feature close to 0.35 cycles/period. Examination of the longitude–time

diagram in Figure 3-1 shows that this fluctuation affects only subpulses occurring at the longitudes of the two outer components. A series of spectra computed separately for different longitudes throughout the profile of this pulsar are shown in Figure 3-4. The feature at 0.35 cycles/period is clearly confined to components 1 and 5, whereas the spectra for components 2 and 4 are essentially featureless. For the region about the profile center including component 3, the spectra are dominated by low-frequency features; this component tends to occur in clumps of 5–10

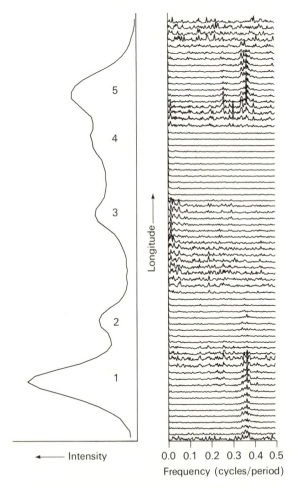

3-4 Separately computed fluctuation spectra for different
longitudes in the profile of PSR 1237 + 25, and the
corresponding integrated profile showing the five
distinct components. [From Backer, 1973.]

subpulses every 20–50 periods (see Figure 3-1). The high degree of symmetry of the fluctuation characteristics about the profile center in this pulsar is striking. Similar symmetries are also seen in other multiple-component pulsars, such as PSR 1133 + 16 and PSR 2045 − 16. In contrast, the fluctuation spectra for PSR 2020 + 28 are not symmetric about the pulse center; the leading component has a weak feature at about 0.36 cycles/period, whereas the trailing component has a strong periodic modulation at 0.47 cycles/period. As discussed in Chapter 2, polarization data for this pulsar suggest that these two components may be emitted from different regions.

A relationship between these periodic modulations and mode changing has been found for PSR 1237 + 25 by Taylor, Manchester, and Huguenin (1975). The strong periodic modulation in component 1 is present only when the pulsar is in its normal mode; in the other mode there may be a weak feature at about 0.24 cycles/period, but the 0.35 cycle/period modulation is completely absent.

The degree of modulation of pulse intensities is best represented by the modulation index, m, defined by

$$m = \frac{(\sigma_{on}^2 - \sigma_{off}^2)^{1/2}}{\langle I \rangle} \tag{3-1}$$

where σ_{on} is the r.m.s. variation of pulse intensities about the mean value $\langle I \rangle$ and σ_{off} is the r.m.s. value of the random noise off the pulse. Observed values of the modulation index (after removing the effects of interstellar scintillation) range between about 0.5 and 2.5 for different pulsars. The modulation is almost invariably deeper at lower radio frequencies. For example, at 147 MHz the modulation index for PSR 0329 + 54 is 2.3, whereas at 400 MHz it is only 1.0. Like the fluctuation spectra, the modulation indexes are different for different longitudes in a given pulsar. Modulation indexes at 400 MHz are plotted as a function of longitude in Figure 3-5 for 12 pulsars. For PSR 1237 + 25 the modulation indexes for the different components are clearly different, and the mirror symmetry seen in the fluctuation spectra is also evident here. For other pulsars changes in modulation index occur within apparently single components. For example, for PSR 1642 − 03 the index is high for the leading half of the profile and low for the trailing half. The modulation index is often higher in the wings of a profile than in the center; examples of this in Figure 3-5 are PSR 1133 + 16, PSR 1929 + 10, and PSR 2016 + 28.

For the highly modulated pulsars, histograms of the pulse intensity have an approximately exponential form, with a maximum at zero and a few pulses with intensity as much as ten times the mean value. For the less modulated sources, histograms usually peak just below the mean

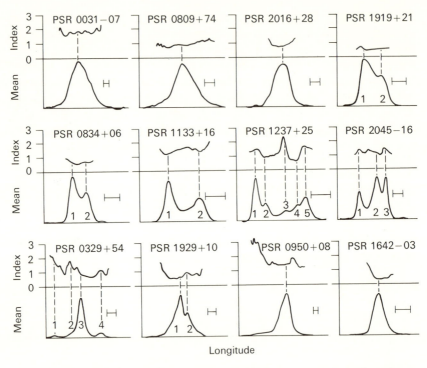

3-5 Modulation indexes at 400 MHz plotted as a function of longitude for twelve pulsars. The longitude scale varies with each graph; horizontal bars represent 5° of longitude in each case. The mean pulse profile is shown for each pulsar, with numbers indicating the different components. [From Taylor *et al.*, 1975.]

value and high-intensity tails do not extend significantly beyond four times the mean energy. Pulsars with extended nulls, e.g., PSR 0031 − 07, often have a bimodal distribution with one of the peaks at zero intensity (Hesse and Wielebinski, 1974; Ritchings, 1976).

Drifting Subpulses

For most pulsars the longitudes of the subpulses in a given pulse are not closely related to those in previous or subsequent pulses. However, in certain pulsars, subpulses in successive pulses drift systematically across the profile. This effect, known as drifting subpulses, was first observed in PSR 1919 + 21 and PSR 2016 + 28 by Drake and Craft (1968). Pulsars exhibiting this phenomenon have been classed as Type D by Taylor and Huguenin (1971). Longitude–time diagrams for three pulsars with prom-

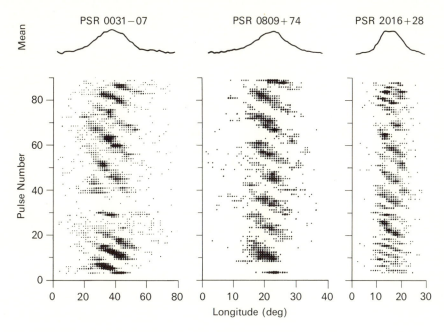

3-6 Longitude–time diagrams for three pulsars with well-organized drifting subpulses. Time increases upward and to the right. [After Taylor *et al.*, 1975.]

inent drifting subpulses are given in Figure 3-6. Subpulses in these pulsars first appear at the trailing edge of the profile and then drift toward the leading edge, disappearing several periods later. A given band of subpulses is usually straight, that is, the amount of subpulse drift per period is constant. Subpulses from two adjacent bands are commonly found within a given individual pulse; the spacing between these subpulses is known as the *secondary period*, P_2. In most Type D pulsars the drifting process results in a strong feature in the fluctuation spectrum; this third period, P_3, is commonly known as the *band spacing*. These parameters are schematically illustrated in Figure 3-7.

The band slope, $|D_\phi| = P_2/P_3$, is quite stable for PSR 0809 + 74, but for PSR 0031 − 07 and PSR 2016 + 28 it varies considerably from one band to another. However, P_2, the spacing between subpulses, tends to remain constant so that when P_3 is large, D_ϕ is small. For PSR 0031 − 07, P_3 generally has one of three approximately harmonically related values, the central one (∼7 pulse periods) being the most common (Huguenin *et al.*, 1970). For PSR 2016 + 28, on the other hand, P_3 ranges between about 3 and 15 pulse periods, with no clearly preferred values.

As shown in Figure 3-3, the fluctuation spectrum for PSR 0943 + 10 has a strong feature close to 0.47 cycles/period. This modulation is

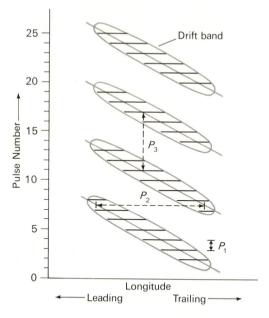

3-7 Schematic longitude–time diagram showing the three periods commonly used to describe drifting subpulses: P_1, the basic pulse period; P_2, the spacing between subpulses; and P_3, the spacing between bands.

probably associated with the clear drifting subpulses seen on longitude–time plots for the pulsar (e.g., Backer, 1973). As for PSR 0809+74, the period corresponding to this modulation frequency is identified with the band spacing, P_3. However, there is an ambiguity in determining P_3 because of possible aliasing effects. The only certain method of identifying the correct drift path is by observing continuity in either the microstructure or the subpulse intensities down one of the possible bands. Unfortunately, such continuity has not been observed with any certainty except for one pulsar, PSR 2016 + 28. An apparently correlated microstructure in successive subpulses from this pulsar was observed by Backer (1973). Further analysis by Cordes (1975b) showed that the correlation is not always present, but that quasiperiodic fluctuations in microstructure are occasionally weakly correlated down a drift band. While it is unlikely that the usually identified drift path is incorrect for such pulsars as PSR 0031 − 07, PSR 0809 + 74, and PSR 2016 + 28, there is considerably more uncertainty for such pulsars as PSR 0943 + 10 for which the modulation frequency is close to the Nyquist value. For this pulsar, modulation at 0.47 cycles/period corresponds to a drift in the direction from the leading to the trailing edge of the profile, i.e., opposite to that for PSR 0809 + 74, etc., at a rate of 4 degrees/period (Sieber and Oster, 1975). If aliasing has occurred and the true frequency is 0.53 cycles/period, then the drift is in the "normal" direction at a rate of 4.4 degrees/period. Other higher-order aliases are of course possible. A similar ambiguity exists in the identification of drift directions for PSR 2303 + 30. For this pulsar an additional complication is introduced by the apparent presence of two distinct drifting systems with different values of P_2 and different modulation frequencies.

Another pulsar having rather complex pulse modulation characteristics is PSR 1944 + 17. As mentioned previously, this pulsar is in a null state for much of the time, emitting short groups of 10 to 20 pulses separated by gaps of 50 to 100 periods. Backer, Rankin, and Campbell (1975) find that in at least some, but not all, of these groups, subpulses drift linearly from the trailing to the leading edge of the profile with a P_3 of about 20 periods.

Parameters for these and other pulsars with relatively well-defined drift behavior are summarized in Table 3-1.

TABLE 3-1

Parameters for pulsars with drifting subpulses

PSR	P_1 (s)	P_2 (deg)	P_3 (periods)	D_ϕ (deg/period)	Ref.[†]
0031 − 07 (A)	0.943	21	13	− 1.6*	1
0031 − 07 (B)	0.943	21	7	− 3.0	1
0031 − 07 (C)	0.943	21	4	− 5.3	1
0301 + 19	1.387	6.2	6.4	− 0.95	2
0809 + 74	1.292	14	11.0	− 1.25	3, 4
0943 + 10	1.097	8.4	2.11 or 1.90	+ 4.0 or − 4.4	5
1919 + 21	1.337	4	4.5	− 0.9	6
1944 + 17	0.440	17	20	− 0.85	3
2016 + 28	0.558	7	3 to 15	− 3 to − 0.5	4
2303 + 30 (A)	1.575	4.2	2.10 or 1.91	+ 2.0 or − 2.2	5
2303 + 30 (B)	1.575	3.4	2.3 or 1.8	− 1.5 or + 1.9	5

* Negative values of D_ϕ represent drift from the trailing to the leading edge of the integrated profile.

[†]REFERENCES: **1.** Huguenin *et al.*, 1970. **2.** Schönhardt and Sieber, 1973. **3.** Backer *et al.*, 1975. **4.** Taylor *et al.*, 1975. **5.** Sieber and Oster, 1975. **6.** Backer, 1973.

For PSR $0031-07$ and PSR $0809+74$, P_2 varies with frequency in the same way as the component separation in multiple-component pulsars, that is, approximately as $v^{-0.25}$ (Taylor *et al.*, 1975). This frequency dependence, which is probably present in other pulsars with drifting subpulses, shows that the origin of drifting subpulses must be closely related to the normal emission processes in pulsars.

Subpulse drift is prominent in the pulsars discussed so far, and in most cases dominates the individual pulse morphology. In many other pulsars, however, more subtle and/or less regular drift is observed. For example, the longitude–time diagram for PSR $1133+16$ (Figure 3-1) shows several groups of subpulses drifting to later longitudes, especially in the trailing half of the profile. This type of behavior can be detected by computing the cross-correlations of successive individual pulses. If drifting behavior is present, the peak correlation coefficient occurs at a delay of slightly less or slightly more than one pulse period, depending on the direction of drift. Results of such an analysis for 15 pulsars are shown in Figure 3-8. In this figure strong drift toward earlier longitudes is indicated for PSRs $0031-07$, $0809+74$, and $2016+28$. Examples of pulsars showing a tendency for subpulses to drift toward later longitudes are PSRs $0834+06$, $1133+16$, and $2021+51$; however, among pulsars in this group the drift is not very regular. Other pulsars, e.g., PSRs $0525 + 21$, $1237 + 25$, and $1642 - 03$, show no evidence of drifting subpulses. Pulsars showing prominent drift toward earlier longitudes almost always have single-peak profiles, and thus are Type SD. They also have relatively small period derivatives. On the other hand, pulsars with subpulse drift toward later longitudes are often Type CD and have moderately large period derivatives. (These correlations are discussed further on pp. 203ff.)

Drifting subpulses can also be detected by analysis of the phase of Fourier components at the frequency of a line feature in the fluctuation spectrum. Using this technique, Backer (1973) has shown that the drift rate of subpulses from PSR $1919+21$ is not constant across the pulse profile. Near the profile center the drift rate (D_ϕ) is about $0.75°$ per period, whereas near the profile edges it is closer to $1.3°$ per period; the value quoted in Table 3-1 is an average value for the entire profile. In an analysis of the fluctuation spectra for this pulsar, Cordes (1975a) has divided the pulse energy into three parts: a steady or slowly varying term, drifting subpulses, and white noise or microstructure. The results show that the drifting-subpulse contribution is greatest at the longitude of the first component and at low frequencies. Relative intensities of the three contributions (integrated over longitude) are shown as a function of frequency in Figure 3-9. The variation with frequency suggests that the white-noise component is associated with the drifting subpulses rather than with the steady component. The drifting subpulses are reduced in

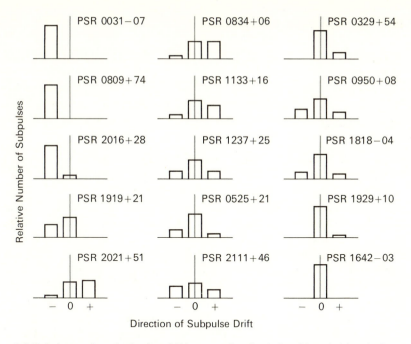

3-8 Relative number of subpulses drifting toward earlier ($-$) and later ($+$) longitudes, and those not drifting (0), based on a cross-correlation analysis. [From Taylor *et al.*, 1975.]

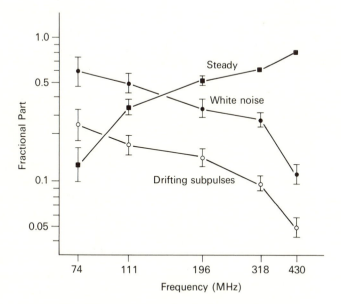

3-9 Frequency dependence of the relative contributions of three terms in the pulse energy of PSR 1919+21: drifting subpulses, white noise, and the steady or slowly varying term. [After Cordes, 1975a.]

intensity when they occur at the longitude of the notch in the integrated profile (described earlier, p. 17).

Micropulses

Early observations by Craft, Comella, and Drake (1968) showed that PSR 0950+08 and PSR 1133+16 exhibit significant variations in pulse intensity on time scales of the order of a few hundred microseconds. These variations, which have time scales smaller than those characterizing subpulses, are known as *micropulses*. As mentioned in Chapter 2, high time resolution can usually be obtained only if some form of dispersion-removal process is employed. Using predetection sampling with a signal bandwidth of about 125 kHz and off-line computer analysis, Hankins (1971, 1972) has been able to investigate pulse structures as short as the inverse of the bandwidth, that is, about 8 μs. In observations at frequencies of 111.5, 196.5, and 318 MHz, Hankins showed that the pulse intensities of PSR 0950+08 and PSR 1133+16 are strongly modulated on time scales down to the resolution limit. An example of a pulse recorded with high time resolution is shown in Figure 3-10.

In many ways the relationship of micropulses to subpulses is similar to the relationship of subpulses to integrated profiles. For example, micropulses generally occur at random positions within subpulses just as subpulses generally occur at random positions within integrated profiles. The pulse illustrated in Figure 3-10 is somewhat atypical in that the microstructure appears to have a quasi-periodic modulation. A similar quasi-periodic microstructure has been observed for PSR 1133 + 16

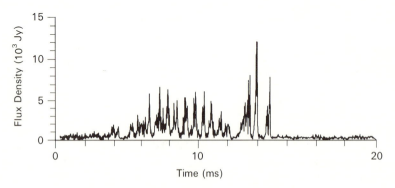

3-10 A de-dispersed pulse from PSR 0950+08 recorded at 111.5 MHz and smoothed to a time resolution of 28 μs. [After Hankins, 1971.]

TABLE 3-2

Micropulse timescales. Only the eight
pulsars listed have been investigated
[*From Cordes, 1975c.*]

PSR	Period (s)	Micropulse timescale, τ_μ	
		(μs)	(deg)
1929 + 10	0.227	None	
0950 + 08	0.253	175	0.25
2020 + 28	0.343	None	
2016 + 28	0.558	290	0.19
1133 + 16	1.188	575	0.17
0834 + 06	1.274	1050	0.30
1919 + 21	1.337	1220	0.33
1237 + 25	1.382	None	

(Ferguson *et al.*, 1976). Of a total of eight pulsars analyzed (Cordes, 1975c), five exhibit microstructure and three do not (Table 3-2).

A useful way of quantifying the presence of microstructure is to compute the average autocorrelation function (ACF) of a number of individual pulses. The ACF typically has the form illustrated in Figure 3-11, with identifiable segments resulting from receiver noise, microstructure, sub-pulse structure, and the integrated profile itself. The lag at the intersection of the microstructure and subpulse segments in Figure 3-11 defines the microstructure time scale, τ_μ, given in Table 3-2. The time scales are independent of frequency for all pulsars observed, at least over the range 111.5 to 318 MHz. The times are approximately proportional to pulsar period, ranging from 0.17 to 0.33 degrees of pulsar longitude.

In a detailed analysis of the pulse structure of PSR 2016 + 28, Cordes (1975b) has shown that autocorrelation functions of this pulsar have the general form shown in Figure 3-11 but with two additional features: (1) a secondary microstructure feature at a lag of 900 $\pm$ 300 μs, and (2) a second subpulse feature at the lag corresponding to the drifting-subpulse period P_2. The first of these features shows that the microstructure for this pulsar has a quasi-periodic variation. As mentioned in the previous section, this modulation is sometimes weakly correlated in several successive pulses of one drift band. The micropulses themselves, however, are not correlated in successive periods.

The frequency dependence of microstructure has been investigated by Rickett, Hankins, and Cordes (1975) by dividing the observed bandwidth

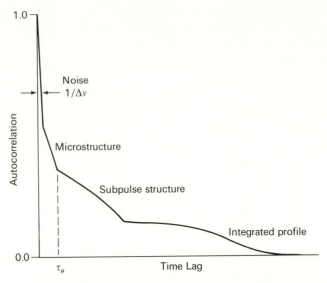

3-11 Typical form of the autocorrelation function for pulses after dispersion removal. The spike at zero lag has a characteristic width equal to the inverse of the sampled bandwidth and represents noise fluctuations. [After Rickett, 1975.]

into several independent bands; this of course involves a corresponding sacrifice in time resolution. Micropulses from PSR 0950+08 were shown to have spectra that are deeply modulated in an apparently random way. These modulations have typical widths of about 6 kHz, equal to the inverse of the microstructure time scale of 175 μs. Cross-correlation of data recorded simultaneously at 111.5 and 318 MHz shows that the microstructure is correlated over this frequency range, with an average correlation coefficient of 0.5 ± 0.2. The micropulses of this pulsar are therefore broadband and cannot result from any frequency-dependent interference effect. The correlation implies that energy densities in the emission region are very large (Chapter 10), and also makes possible extremely accurate measurements of dispersive delay.

Polarization

For pulsars in which the integrated profile is highly polarized, essentially all subpulses must likewise be highly polarized and have stable polarization characteristics. However, many pulsars have rather weakly polarized integrated profiles. Three possible reasons for low polarization are that subpulses at a given longitude may (1) be themselves weakly polarized; (2) be divisible into groups with orthogonal polarization; or (3) have randomly varying position angles and sense of circular polarization. Observations described in this section show that all three effects are seen in the radio emission from pulsars.

Polarization characteristics for a series of consecutive pulses from PSR 0329 + 54 are given in Figure 3-12. Many subpulses from this pulsar are

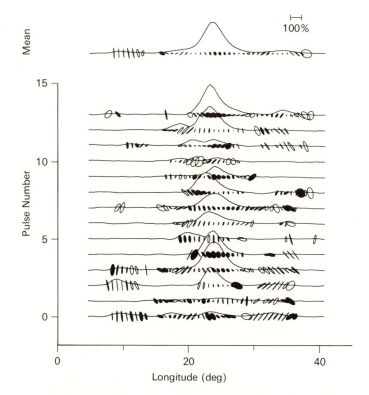

3-12 Polarization characteristics of 14 consecutive pulses from PSR 0329 + 54, recorded at a frequency of 410 MHz, and of the integrated profile obtained by summing these pulses. The state of polarization of the signal at each sample time is represented by an ellipse plotted under a line representing total intensity of the pulse. Filled ellipses indicate left-circular polarization, open ellipses right-circular polarization. The major axis of each ellipse is proportional to the degree of polarization (the bar in the upper right corner measures the major axis length corresponding to complete polarization). Each ellipse is oriented according to position angle (with arbitrary zero). [After Manchester *et al.*, 1975.]

highly polarized, with linear polarization generally dominating over circular. The variation of position angle across subpulses is usually rather small, typically less than 30°. These characteristics are also typical for other pulsars.

In pulsars with multiple-component profiles the degree of polarization of the subpulses generally varies considerably from one component to another. For example, in PSR 0329 + 54 component 1 is relatively highly polarized (Figure 3-12; and see Figure 2-10, p. 29), so subpulses are also highly polarized and have stable position angles. On the other hand, the main component (3) is less than 20 percent polarized. The intensity of subpulses falling at the longitude of component 3 is plotted against their percentage polarization in Figure 3-13. This figure shows that the main reason for the rather low integrated polarization is that most subpulses are themselves weakly polarized. There is also some scatter in the position angle of these subpulses (Manchester *et al.*, 1975), which further reduces the polarization of the integrated profile.

In many pulsars, particularly at lower frequencies, the position angle of subpulses at a given longitude is quite stable, but there is a wide variation in their fractional polarization. The fractional polarization of the integrated profile is then equal to the mean polarization of the subpulses, weighted by their intensities. For example, in PSR 1237 + 25 almost all subpulses have the same position angle, so that the position angle is almost constant across the integrated profile. However, the subpulses

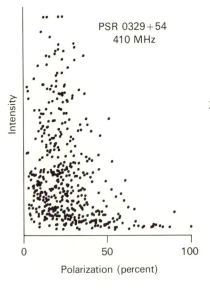

3-13 Intensity (on a linear scale) versus percentage polarization of subpulses falling at the longitude of the main component of PSR 0329 + 54, recorded at a frequency of 410 MHz. [After Manchester *et al.*, 1975.]

can be divided into two groups: those with essentially complete linear polarization and those that are only partially polarized. For components 2 and 4 of the integrated profile (see Figure 2-12, p. 31) about 30 percent of the subpulses are fully polarized and the peak of the polarization distribution for the remainder falls at about 85 percent, whereas for components 1 and 5 only 15 percent are fully polarized and the peak of the distribution for the other group is near 70 percent polarization. Very few subpulses at the longitude of component 3 are fully polarized. Consequently, the integrated polarization is quite high for components 2 and 4, somewhat lower for components 1 and 5, and lowest for component 3.

Another way in which integrated profiles are depolarized is by the occurrence of orthogonally polarized subpulses. For many pulsars the majority of subpulses at a particular longitude are polarized with approximately the same position angle, but occasional subpulses are polarized at the orthogonal position angle. These orthogonal subpulses were seen in essentially all of the pulsars studied by Manchester, Taylor, and Huguenin (1975), so the effect is relatively common. In general, the proportion of subpulses in the orthogonal polarization mode varies with longitude across the profile—in some cases the orthogonal mode may become dominant, leading to a 90° discontinuity in the integrated profile position angle (c.f. Figure 2-10, p. 29), Orthogonal subpulses are observed in the fourth (trailing) component for PSR 0329 + 54 (Figure 3-12); near the peak of the integrated profile for PSR 0950 + 08; in the leading component for PSR 1133 + 16; in the trailing half of the PSR 2021 + 51 profile; and in the wings of the PSR 2045 − 16 profile. Distributions of subpulse position angles for these pulsars at these longitudes are shown in Figure 3-14. The distributions are clearly bimodal, and in each case the difference in position angles between the two groups of subpulses is consistent with orthogonality. Apart from this difference, orthogonal subpulses generally have properties very similar to normal subpulses. For example, distributions of both subpulse intensity and fractional polarization are usually similar for the two groups. The occurrence of orthogonal subpulses in a given component seems to be random in time and not related in any obvious way to the occurrence of normal subpulses.

In PSR 1929 + 10, which has a very highly polarized integrated profile (Figure 2-9, p. 27), about two percent of the subpulses observed in the trailing part of the profile are orthogonal. These orthogonal subpulses are usually superimposed on the normal emission, resulting in a rather weakly polarized pulse of greater than average intensity. The extreme wings of the integrated profiles of Type C pulsars, such as PSR 1237 + 25 and PSR 2045 − 16, are weakly polarized (p. 25). This low polarization results from the occurrence of orthogonal subpulses at these longitudes.

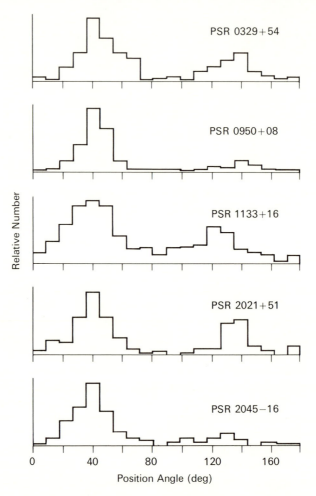

3-14 Distribution of subpulse position angles for five pulsars at the longitudes at which orthogonal subpulses are seen. Several hundred pulses were observed for each of the pulsars. [From Manchester *et al.*, 1975.]

The third way in which integrated profiles are depolarized is illustrated by a comparison of the high- and low-frequency polarization characteristics of PSR 0950+08. Figure 3-15 plots the percentage polarization versus the position angles of subpulses occurring at the trailing edge of the integrated profile from a sequence of 500 pulses at each of two frequencies. At the lower frequency, 410 MHz, all subpulses have essentially

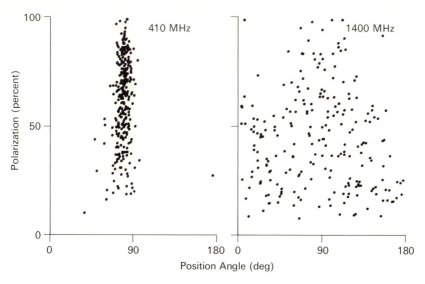

3-15 Percentage linear polarization plotted against position angle for subpulses at the trailing edge of the integrated profile of PSR 0950+08. [From Manchester *et al.*, 1975.]

the same position angle, but there is a wide scatter in the degree of polarization; the polarization of this part of the integrated profile is relatively high, about 55 percent. For the same longitude range at 1,400 MHz, the distribution of fractional polarization is essentially the same as at the lower frequency, but the position angles are almost randomly distributed; the scatter of position angles reduces the polarization of the integrated profile at this frequency to about 10 percent. A similar decrease in stability of position angles with increasing frequency is seen in other pulsars; this effect is the primary reason for the observed high-frequency depolarization of integrated profiles (see Figure 2-9, p. 27). In most cases there is also some reduction in the average polarization of subpulses at higher frequencies, which further reduces the polarization of the integrated profile.

Figure 3-16 shows the polarization characteristics of a consecutive series of pulses from PSR 0809+74, a pulsar with regularly drifting subpulses. The integrated profiles of this and other Type D pulsars are weakly polarized, although individual subpulses are often quite highly polarized. As may be seen in Figure 3-16, the variation in position angle across most subpulses is similar—the total variation across a given subpulse is often close to 90°. The position angle at a given point within a

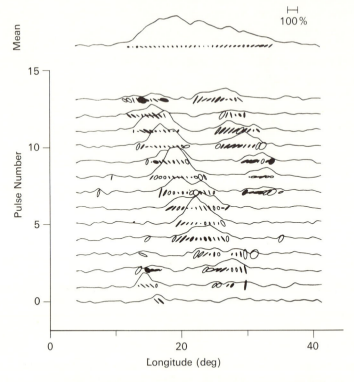

3-16 Polarization characteristics for a consecutive sequence of pulses from PSR 0809 + 74, recorded at a frequency of 147 MHz. The state of polarization of the signal at each longitude is represented by an ellipse (see Figure 3-12 for explanation). [After Manchester *et al.*, 1975.]

subpulse, say the trailing edge, is approximately constant from one sub-pulse to another; that is, position angle variations are related to the subpulses rather than to the integrated profile. This fact, together with the regular drift of the subpulses, accounts for the low polarization of the integrated profile of PSR 0809 + 74. By averaging subpulses from a number of drift bands of this pulsar, Manchester, Taylor, and Huguenin (1975) showed that the position angle does not vary smoothly across subpulses, as it does in most other pulsars, but changes discontinuously by about 90° near the subpulse peak. The location of the position angle transition relative to the subpulse peak appears to vary as the subpulse drifts across the profile. When the subpulse is near the center of the in-tegrated profile, the transition occurs about one degree of longitude before the subpulse peak; the transition occurs somewhat earlier when the subpulse is near the leading edge of the profile, and somewhat later when the subpulse is near the trailing edge of the profile. Similar 90° transitions of position angle are observed in other Type D pulsars, such as PSR 0031−07 and PSR 2016+28. Backer, Rankin, and Campbell (1976) suggest that in PSR 2016+28 there are two 90° position angle transitions in each subpulse, one near the leading edge of the subpulse and one near the trailing edge. Again, the location of these transitions appears to change as the subpulse drifts across the profile.

Figures 3-12 and 3-16 show that subpulses sometimes have a high degree of circular polarization. In most pulsars the intensity and sign of this circular polarization changes in an apparently random way, leading to only a small degree of circular polarization in the integrated profile. However, the integrated profiles of some pulsars have stronger circular polarization (see Figure 2-11, p. 29). For these pulsars, the circular polarization appears to be related to the integrated profile, rather than to individual subpulses as in most pulsars. For example, subpulses falling at the longitude of component 3 in PSR 1237+25, generally have left-circular polarization on their leading edge and right-circular polarization on their trailing edge. However, the change of sign precedes the subpulse peak by about 2 ms, falling very close to the midpoint of the integrated profile.

Circular polarization associated with subpulses rather than the integrated profile has been observed in the Type D pulsar PSR 2303+30 by Rankin, Campbell, and Backer (1974). The detailed behavior is complex, but at times a sense reversal appears to occur close to the center of the subpulse and to move with the subpulse as it drifts through the integrated profile. No significant circular polarization associated with subpulses was observed by Manchester, Taylor, and Huguenin (1975) for PSR 0031−07 and PSR 0809+74.

Observations by Cordes (1975c) have shown that in PSRs 0950+08, 1133+16, 1919+21, and 2016+28 the polarization variations associated with micropulses resemble those associated with subpulses. In general, the degree of polarization and the position angle tend to remain constant within a micropulse, although autocorrelation analysis indicates some variation in micropulses from PSR 0950+08. Adjacent micropulses in pulsars with highly polarized integrated profiles have the same or similar polarization properties. However, in other pulsars micropulse polarization is greater than the subpulse polarization and there is considerable variation of polarization characteristics between adjacent micropulses. In particular, 90° transitions of position angle between adjacent micropulses are frequently observed. These position angle transitions are usually accompanied by a reversal in the sense of circular polarization, so the two polarization states are truly orthogonal, i.e., they are separated

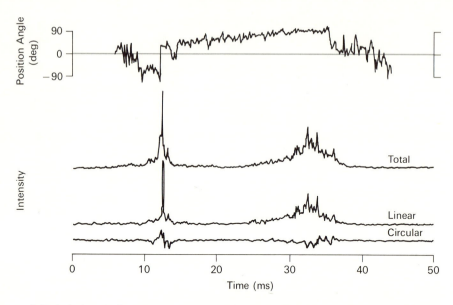

3-17 A single pulse from PSR 1133 + 16 recorded at 430 MHz, plotted with 128 μs resolution. The four curves represent the total intensity, the linearly and circularly polarized components, and the position angle. Two 90° jumps in position angle are evident: one near the leading edge of the strong micropulse in the first subpulse, and a slower one near the trailing edge of the second subpulse. [After Cordes, 1975c.]

by 180° on the Poincaré sphere. These characteristics are illustrated in an individual pulse in Figure 3-17. Orthogonal transitions between micropulses may be responsible for the transitions seen within subpulses in Type D pulsars, and occasionally in pulsars of other types.

4

The Crab Nebula
and Its Pulsar

The Crab Nebula, remnant of a supernova explosion observed by the Chinese as a "guest star" in 1054 A.D., occupies a unique place in astronomy. Given the name Crab by Lord Rosse in 1844, the nebula was first associated with the 1054 A.D. event by Hubble in 1928 and conclusively identified with it by Duyvendak, Oort, and Mayall in 1942. Photographs of the nebula taken in several colors by Baade (1942) showed that it consists of a network of line-emitting filaments superimposed upon a more uniform region of continuous emission (Figure 4-1). All but a few percent of the visible light comes from this continuum. The radio source Taurus A was discovered by Bolton in 1948 and identified with the Crab Nebula by Bolton, Stanley, and Slee (1949)—the first radio source to be identified with an optical object. In 1953 Shklovskii proposed that the optical continuum radiation was generated by the synchrotron process; his prediction that the light from the nebula should be linearly polarized was confirmed the following year. A strong source of X-rays in Taurus was identified with the Crab Nebula by Bowyer and co-workers (1964) in one of the first identifications of a celestial X-ray source. At about the same time a point-like low-frequency radio source was discovered near the center of the nebula by Hewish and Okoye (1964). Finally, of course,

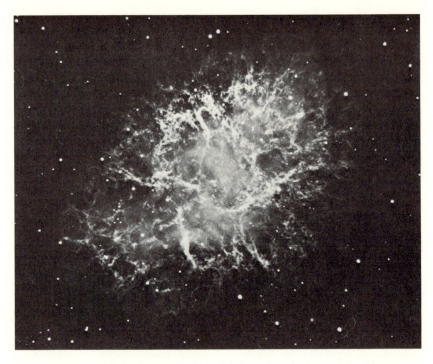

4-1 The Crab Nebula photographed in red light to show the line-emitting filaments.
[Courtesy of the Lick Observatory.]

a pulsar was found in the nebula by Staelin and Reifenstein (1968). It has the shortest period of any known pulsar and has been shown to be ultimately responsible for most of the emission from the nebula, including the X-rays and the compact radio source. Because of this close association between the nebula and the pulsar, we begin this chapter with a description of the nebula itself.

The Nebular Emission

At radio and X-ray frequencies the Crab Nebula is one of the strongest sources in the sky, but at optical frequencies it is not so prominent. Its apparent visual magnitude is about 8.4 and, although its surface brightness is rather low, it can be seen with a small telescope. Continuum spectra from low radio frequencies to the γ-ray region for both the nebula and the pulsar are given in Figure 4-2. The nebular spectrum appears to be essentially continuous over this entire range, although observations are lacking

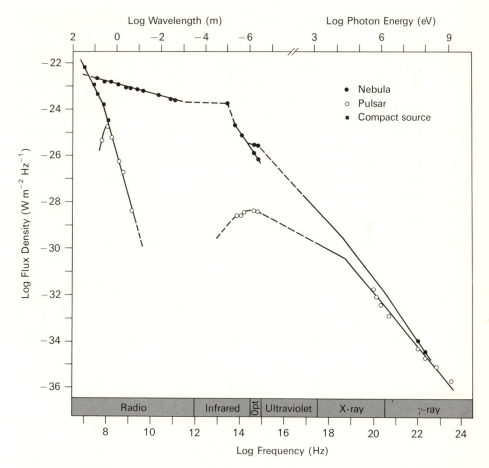

4-2 Spectra of the Crab Nebula and its pulsar over the range 10^7 to 10^{24} Hz. Two lines are drawn for the nebular spectrum in the optical region: the lower line is the observed spectrum and the upper line is corrected for interstellar extinction ($A_v = 1.6$ mag). The optical spectrum for the pulsar has also been corrected for $A_v = 1.6$ mag. Data from the following sources were used: Aitken and Polden, 1971; Andrew et al., 1964; Becklin and Kleinmann, 1968; Becklin et al., 1973; Bell and Hewish, 1967; Bridle, 1970; Clark et al., 1973; Kellermann et al., 1969; Kniffen et al., 1974; Kurfess, 1971; Laros et al., 1973; Manchester, 1971b; Matveyenko, 1971; Matveyenko and Meeks, 1972; McBreen et al., 1973; O'Dell, 1962; Oke, 1969; Parker, 1968; Rankin et al., 1970; Scargle, 1969; Thomas and Rothenflug, 1974; Vandenberg et al., 1973; and Williams et al., 1965.

in the far infrared and ultraviolet regions. With the possible exception of the highest γ-ray frequencies, the nebular emission is almost certainly generated by the synchrotron process. The luminosity derived from the nebular spectrum (assuming isotropic emission) is 1.0×10^{38} erg s^{-1}, of which about 12 percent is emitted at radio frequencies.

High-resolution maps at radio frequencies (Figure 4-3) show that the nebula is roughly elliptical in shape with a half-intensity width of 3.5 arc for the major axis (at position angle 135°) and 2.3 arc for the minor axis. For a distance to the nebula of 2 kpc (see below), the corresponding linear dimensions are approximately 2.0 pc and 1.3 pc. At optical frequencies the continuum emission has half-intensity widths of about 2.5 and 1.5 for the major and minor axes. Although the optical half-intensity widths

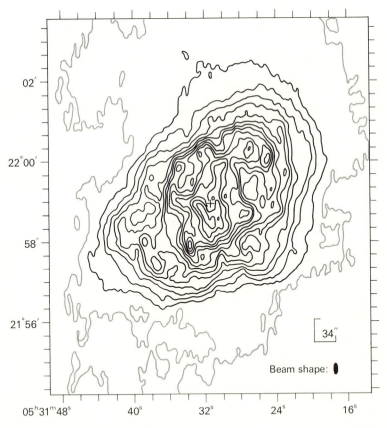

4-3 Distribution of radio emission of the Crab Nebula at 5 GHz. The Crab pulsar is located in the center of the map (marked with a cross). [From Wilson, 1972a.]

are smaller, the overall extent of detectable emission is about $7' \times 5'$ at both radio and optical frequencies.

There is considerable structure within the nebula, the most prominent features being the so-called "bays" in the southeast and northwest edges of the nebula. Because of these bays, which can be seen in Figures 4-1 and 4-3, the continuum emission forms an S-shaped ridge along the major axis of the nebula; there is no evidence for the shell structure seen in most supernova remnants. Several of the ridges in Figure 4-3 coincide with line-emitting optical filaments; Wilson (1972b) attributed this increased emission to synchrotron radiation in stronger magnetic fields surrounding the filaments. These stronger fields probably result from currents flowing along the filaments and may confine the filaments by means of the pinch effect. Although the optical continuum is often called "amorphous," the highest resolution photographs show that it consists of a mass of very fine filaments that may be aligned with the nebular magnetic field.

As can be seen in Figure 4-2, possible discontinuities appear in the nebular spectrum in the far infrared and optical regions. Aitken and Polden (1971) suggest that their high 10-micron flux could result from reradiation by dust grains in the nebula. The slope of the optical continuum is very dependent on the rather poorly known interstellar extinction. From observations of the relative intensity of the [SII] lines from the filaments, Miller (1973) obtained a value of 1.6 ± 0.2 magnitudes for A_v, the visual absorption, but the spectral continuity would be better for $A_v \approx 1$ magnitude.

Although these line-emitting filaments contribute only a small fraction of the total optical emission, they contain most of the nebular mass, the total being about $1\ M_\odot$. Davidson and Tucker (1970) suggest that the filaments could have neutral cores and hence even greater mass; however, the total mass of the nebula probably does not exceed $2\ M_\odot$. The angular thickness of the filaments is typically a few arc seconds, so if one assumes a cylindrical cross section their linear diameter must be 0.01–0.05 pc. The principal spectral lines from the filaments are the forbidden lines of [OII] (3726, 3729 Å), [OIII] (4959, 5007 Å), [NII] (6548, 6583 Å) and [SII] (6716, 6731 Å), as well as Hα (6563 Å). It is therefore clear why the filaments are most prominent on photographs taken in the red region of the spectrum (Figure 4-1). Electron densities in the line-emitting regions may be derived from the intensity ratio of the lines in the [OII] doublet—a mean value of about $10^3\ cm^{-3}$ is obtained. The temperature is more difficult to determine but is thought to be about 10^4 K, so conditions in the filaments are in many ways similar to those in the denser galactic HII regions.

At X-ray wavelengths the spectral index is close to -1.2 at the lower energies and probably somewhat steeper at energies around 100 MeV. By comparison, the radio frequency spectrum is much flatter, with a spectral index of −0.26. The low intensity of γ-rays with energies greater than 250 GeV measured by Fazio and co-workers (1972) implies a steepening of both the nebular and pulsar spectra above 1 GeV. The X-ray emitting region is not a point source, as was first demonstrated in a lunar occulation experiment by Bowyer and co-workers (1964). Recent satellite and rocket observations of subsequent lunar occulations by several groups (e.g., Davidson *et al.*, 1975; Ricker *et al.*, 1975) show that the angular diameter of the source is about 70″ arc at energies of a few keV and about half this at energies of the order of 100 keV. The centroid of the emission is displaced about 10–20″ arc west from the position of the pulsar.

The distribution of optical polarization in the nebula is shown in Figure 4-4. In the central regions the light is about 40 percent linearly polarized, with the electric vector approximately parallel to the bright ridge. If the optical emission is generated by the synchrotron process, as is strongly suggested by the nonthermal spectrum and the presence of polarization, the direction of the magnetic field is perpendicular to the

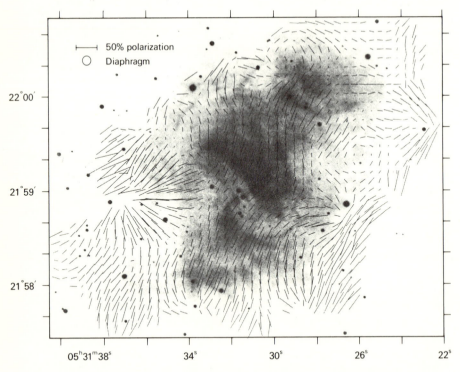

4-4 Linear polarization of the continuum optical emission of the Crab Nebula. The lines are oriented according to the direction of the electric vector and their lengths are proportional to the percentage polarization. [From Woltjer, 1957.]

ridge. In the outer regions the polarization is higher, reaching 60 percent around the southeast bay where the radial configuration of vectors suggests a current flowing parallel to the line of sight.

At radio wavelengths the degree of polarization is less, typically 10–15 percent of the total intensity. Wilson (1972a, 1974) finds that the general distribution of radio polarization is similar to that at optical wavelengths; however, in the southeast bay and some other regions there is little correlation. The observed differences between the optical and radio polarizations can be largely accounted for by Faraday rotation of the radio radiation within the line-emitting filaments. If, within the filaments, the field strength is 5×10^{-4} G and the electron density (n_e) is approximately 10^3 cm^{-3}, then for a path-length of 0.02 pc the rotation measure (see p. 134) would be about 10^4 rad m^{-2}. The radio emission would be effectively depolarized by propagation through a medium characterized by rotation measures as large as this. Rotation measures of the Crab pulsar ($RM = -42.3$ rad m^{-2}) and the nearby object PSR 0525+21 ($RM = -39.6$ rad m^{-2}) are very similar, suggesting that the radiation from the Crab pulsar suffers very little Faraday rotation within the nebula. As magnetic fields within the nebula are not small (see below), this implies that the thermal electron density outside the filaments is less than about 0.01 cm^{-3}.

Circular polarization in the radio emission from the nebula has been detected by Weiler (1975). A synthesis map at 1415 MHz reveals several regions with about 0.05 percent right-circular polarization; no left-circular polarization is detected. This compares with upper limits of about 0.03 percent placed on the circular polarization of the nebular optical continuum by Landstreet and Angel (1971). At X-ray wavelengths, Weisskopf and co-workers (1976) have found that the integrated linear polarization of the nebular radiation is about 15 percent, with a position angle similar to that of the optical emission in the central region. This is strong evidence that synchrotron emission is responsible for the nebular X-rays as well as the radio and optical emission. As X-rays from other supernova remnants are most probably generated by the thermal bremsstrahlung process, the Crab Nebula appears to be unique in this respect.

A relativistic electron (of charge e and rest mass m) in a magnetic field generates synchrotron emission with maximum intensity at a frequency

$$\nu_m \approx 0.07 \frac{eB_\perp}{mc} \left(\frac{\varepsilon}{mc^2} \right)^2 \approx 1.8 \times 10^{18} B_\perp \varepsilon^2 \quad \text{(Hz)}, \qquad (4\text{-}1)$$

where $B_\perp$ is the component of the magnetic field perpendicular to the electron velocity (in gauss, G) and ε is the electron energy (in ergs). Adopting a value for the magnetic field strength of 5×10^{-4} G (see below), the electron energy required for radio radiation at 10^7 Hz is 10^{-4} erg or 70 MeV; for γ-rays at 10^{22} Hz it is 2×10^{15} eV. Gould (1965) has shown that inverse Compton scattering of the radio and X-ray photons by these high-energy electrons produces high-energy γ-rays. The observed γ-rays at frequencies greater than 10^{22} Hz are probably generated by this process.

The power radiated by an electron is given by

$$-\frac{d\varepsilon}{dt} = \frac{2}{3} c \left(\frac{e^2}{mc^2}\right)^2 \left(\frac{\varepsilon}{mc^2}\right)^2 B_\perp^2 \approx 2.4 \times 10^{-3} B_\perp^2 \varepsilon^2 \quad (\text{erg s}^{-1}). \quad (4\text{-}2)$$

For an ensemble of electrons with a differential energy distribution

$$N(\varepsilon)\, d\varepsilon = K\varepsilon^{-\gamma}\, d\varepsilon, \quad (4\text{-}3)$$

where K and γ are constants, the specific intensity (for an optically thin source) is

$$I_\nu = A(\gamma) K l B^{(\gamma+1)/2} \nu^{-(\gamma-1)/2}, \quad (4\text{-}4)$$

where $A(\gamma)$ is a function of γ and l is the source dimension (Ginzburg and Syrovatskii, 1965). In the radio range the nebular spectral index is -0.26, so $\gamma = 1.52$. For a magnetic-field strength of 5×10^{-4} G, the density of relativistic electrons required to produce the observed radio radiation is about 10^{-5} cm^{-3}. The contribution of these electrons to the energy content of the nebula is $\sim 2 \times 10^{47}$ ergs. Because of the increased efficiency of the radiation process at high energies (Equation 4-2), the relativistic electron density required to produce the optical and X-ray emission is only $\sim 10^{-8}$ cm^{-3}, corresponding to an energy content of about 4×10^{47} ergs.

From Equations 4-1 and 4-2 the lifetime of the relativistic electrons is

$$\tau_R \approx \varepsilon/(-d\varepsilon/dt) \approx 6 \times 10^{11} B^{-3/2} \nu_m^{-1/2} \quad (\text{s}). \quad (4\text{-}5)$$

For radio emission $\nu_m \lesssim 10^{11}$ Hz, so $\tau_R \gtrsim 2 \times 10^{11}$ s or 6,000 years; it is thus possible that the electrons responsible for the radio emission were produced at the time of the supernova. However, for the optical and X-ray emissions the lifetime of the electrons is much less than the age of the nebula; for example for X-rays with $\nu_m \approx 10^{20}$ Hz the electron lifetime is $\tau_R \approx 6 \times 10^6$ s or 10 weeks. Clearly, such electrons cannot have been produced at the time of the supernova and still be radiating. So, if the optical and X-ray emission is to be interpreted as synchrotron emission, continuous injection or acceleration of relativistic electrons is required.

Before the discovery of the Crab pulsar this represented a major problem in understanding the physics of the Crab Nebula. It is likely that pulsars (directly or indirectly) are efficient in accelerating particles to ultra-relativistic energies. The rate of loss of rotational kinetic energy from the Crab pulsar ($\dot{W} = I\Omega\dot{\Omega}$, where I, the moment of inertia, is thought to be 10^{45} gm cm^2) is about 5×10^{38} erg s^{-1}; this is more than adequate to account for the nebular luminosity of 10^{38} erg s^{-1}. At energies for which the lifetime of the relativistic electrons is less than the age of the nebula (greater than 6×10^{10} eV, corresponding to synchrotron emission at $\gtrsim 10^{13}$ Hz), the energy spectrum of the electrons contained in the nebula is steeper than that of the injected electrons because of the ε^2 dependence of the synchrotron loss rate (Equation 4-2). If the energy spectrum of the injected electrons is proportional to $\varepsilon^{-\gamma}$, then at high energies the spectrum of the electrons in the nebula varies as $\varepsilon^{-(\gamma + 1)}$ and the synchrotron spectral index is decreased by 0.5. At infrared and optical frequencies the observed spectrum may be close to a power law with spectral index of -0.8, but at higher frequencies the spectrum is steeper, suggesting that the injected electron spectrum is also steeper at high energies.

The X-ray source at high energies (~ 100 keV) has a diameter of about one light year. Therefore, because of the short lifetime against radiation losses, the radiating electrons cannot be accelerated at the pulsar. Also, Tademaru (1973) and Cocke (1975) have pointed out that, because of curvature radiation losses (see p. 184), electrons injected into the nebula by the pulsar will have an upper energy limit of about 3×10^{12} eV, corresponding to a synchrotron frequency of about 10^{16} Hz. Cocke has proposed that further acceleration occurs by a Fermi process in a small region (radius ~ 0.2 pc) surrounding the pulsar. This model provides a natural explanation for the extended nature of the hard X-ray source and its steep spectrum. The total rate of particle injection from the pulsar required to account for the observed nebular luminosity is about 10^{38} particles per second.

Magnetic Fields in the Nebula

Interpretation of the nebular continuum as synchrotron radiation requires that a relatively strong magnetic field exist throughout the nebula. Several methods of estimating the field strength are available. Since relativistic electrons are contained within the nebula, presumably by the magnetic field, the particle energy density cannot exceed the magnetic field energy density $B^2/8\pi$. However, the field strength and particle energy are also related by the observed synchrotron luminosity (see Equation 4-2). For equipartition (particle and field energy densities equal), the required field strength is about 5×10^{-4} G, corresponding to a total energy content (both particles and field) of about 10^{48} ergs. The combined energy content increases rapidly if the magnetic field is varied either up or down from the equipartition value, so the system would be expected to relax toward equipartition. A second estimate of the field strength can be obtained from the observed steepening of the nebular spectrum at about 10^{13} Hz. From Equation 4-5 the break frequency is given by

$$v_b \approx 4 \times 10^{23} B_\perp^{-3} \tau_R^{-2} \quad \text{(Hz)}, \tag{4-6}$$

where τ_R is in this case the age of the nebula; thus, for $\tau_R = 2.8 \times 10^{10}$ seconds, B $\approx 4 \times 10^{-4}$ G. The interpretation of high-energy γ-rays from the nebula as inverse Compton radiation provides a third estimate of the field strength. To produce the flux observed by Fazio and co-workers (1972), the average field strength within the nebula would have to be close to 10^{-3} G. These various estimates suggest that the average field strength within the nebula is between 5×10^{-4} and 10^{-3} G.

The origin of this field, which the polarization measurements show has a rather regular large-scale structure, is a difficult problem. It cannot have resulted from a simple expansion of a stellar field because the ratio of gravitational to magnetic energy (which is conserved in a spherical expansion) is at present only about 10^{-6}. Consequently, the system could never have been stable. Furthermore, any field presently attached to the pulsar cannot be of sufficient strength; even for an inverse-square dependence on distance, the nebular field from this source would be $< 10^{-12}$ G throughout most of the nebula. A swept-up interstellar field would be confined to the edges of the nebula, giving it a shell structure rather than the observed structure with peak intensity near the center. Nor do turbulent motions within the nebula seem adequate to amplify a seed field to the present observed strength.

It seems therefore that the field must have been generated by the pulsar since the initial explosion. As described in Chapter 9, the pulsar is expected to generate a large flux of magnetic-dipole radiation. Earlier, Rees (1971) suggested that the observed magnetic field in the Crab nebula was simply the field associated with these electromagnetic waves. Because its radial dependence ($\sim r^{-1}$) is much weaker than that of a dipole ($\sim r^{-3}$), this field would have a strength of about 10^{-4} G at the edges of the nebula. Electrons trapped by this magnetic-dipole radiation would have a gyro-frequency much higher than the wave frequency and so emit "synchro-Compton" radiation with characteristics very similar to synchrotron radiation. Since the magnetic-dipole radiation away from the equatorial plane has a circularly polarized component, the synchro-Compton radiation would likewise be circularly polarized with an expected degree of about one percent. As mentioned above, an upper limit of about 0.03 percent has been placed on the degree of circular polarization of the optical continuum, which appears to rule out this interpretation of the field. Subsequently, Rees and Gunn (1974) proposed that the magnetic-dipole radiation is absorbed in a shock discontinuity at a radius of about 0.1 pc from the pulsar, the energy being converted to relativistic particles. Following an earlier suggestion of Piddington (1957), they suggest that, outside the shock, field amplification occurs by "winding-up" of a toroidal field component. This process would be expected to be self-limiting at about the equipartition field strength. However, observations of circular polarization at 1415 MHz by Weiler (1975) seem more consistent with an average field directed toward the earth over the whole nebula than with a toroidal field. In sum, the mechanism by which the nebular magnetic field is generated is not yet clear.

Expansion of the Nebula

Proper motion of the filaments resulting from expansion of the nebula was first detected by Duncan (1939); he also showed that the age of the nebula calculated on the basis of uniform expansion was less than the known age, implying that the filaments had been accelerated since the supernova. In an analysis of plates taken over a 30-year period, Trimble (1968) found that the expansion velocity of the filaments is proportional to their distance from the center. The largest observed velocities, about 1500 km s^{-1}, were observed near the ends of the major axis. By combining radial velocity and proper motion measurements with the assumption that the nebula is a prolate spheroid with its major axis in the plane of the sky, Trimble obtained a distance of 2 kpc for the nebula. The date of convergence of the filaments is about 1140 A.D., or 86 years after the known date of the supernova. This corresponds to a uniform acceleration $\dot{v} \approx 10^{-3}$ cm s^{-2} at the ends of the major axis. The power required to maintain this acceleration is

$$\dot{W} = Mv\dot{v} + \tfrac{1}{2}v^2\dot{M}, \tag{4-7}$$

where M is the mass of the nebula, v is the expansion velocity, and $\dot{M}$ is the rate at which the nebula sweeps up interstellar material. Taking $M = 1\,M_\odot$ and $v = 1.5 \times 10^8$ cm s^{-1}, we obtain

$$\dot{W} = 3 \times 10^{38} + 5 \times 10^{38}\,n_H \quad (\text{erg s}^{-1}), \tag{4-8}$$

where n_H is the neutral hydrogen density in the surrounding interstellar medium. For $n_H \approx 0.2$ cm^{-3}, the total power required is $\sim 4 \times 10^{38}$ erg s^{-1}. It therefore seems that a large fraction of pulsar energy loss is required to maintain the expansion of the nebula. On the basis of the parameters given above, the relativistic particles and magnetic fields within the nebula exert sufficient pressure to cause the observed acceleration.

The early observations by Baade revealed time-variable features not related to the nebular expansion, especially in the central part of the nebula. Scargle (1969) found that the principal activity was in a series of "wisps" lying a few arc seconds northwest of a star now known to be the pulsar. These wisps are elongated regions of enhanced emission, with typical dimensions about $6'' \times 2''$ arc, that vary in intensity on time scales of months to years and appear to move at velocities of up to 0.3 c. The energy associated with each burst of activity is about 10^{43} ergs. Scargle interprets these wisps as local enhancements of the synchrotron emission associated with hydromagnetic activity generated by the pulsar—they are strongly polarized with the implied magnetic field along their length.

The Crab Pulsar

The discovery by Staelin and Reifenstein (1968) of two pulsars near the Crab Nebula, and the subsequent observation by Comella and co-workers (1969) that one of them was within the nebula and had the very short period of 33 ms, added a new dimension to the investigation of the Crab Nebula. With the observation of the regularly increasing period of this pulsar (Richards and Comella, 1969), it became clear that a likely solution to the energy-supply problem for the nebula had been found. The significance of these discoveries was further enhanced when Cocke, Disney, and Taylor (1969) discovered optical pulses with the same period as the radio pulses. Subsequent observations by Lynds, Maran, and Trumbo (1969) showed that these pulses came from the south-preceding member of the pair of 16th-magnitude stars near the center of the nebula—the star identified by Baade and Minkowski in 1942 as the stellar remnant of the supernova. The dramatic television detection of the pulsar by Miller and Wampler (1969) is shown in Figure 4-5. This observation showed that the intensity of the pulsar at minimum light was less than two percent of the maximum intensity, so essentially all the light from this star is in pulsed form. Of course, the rapid pulsing of the light was not obvious in the earlier long photographic exposures. The spectrum of the pulsar was further extended by the discovery of X-ray pulses by Fritz and co-workers (1969) and of γ-ray pulses by Hillier and co-workers (1970).

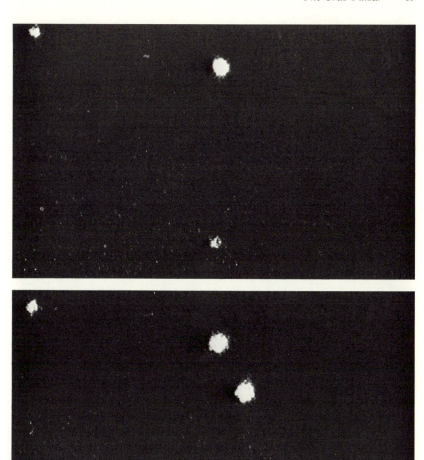

4-5 Television detection of the Crab pulsar. In the upper picture the pulsar (the south-preceding star of the pair) is near minimum light, and in the lower picture near maximum light. [From Miller and Wampler, 1969.]

Integrated profiles of the Crab Nebula pulsar in each of the four frequency regimes (radio through γ-ray) have the same basic shape, with a main pulse and a relatively strong interpulse (Figure 4-6). The separation of the main pulse and interpulse at radio and optical frequencies is 13.37 ± 0.03 ms or 40.4 percent of the period. At radio frequencies

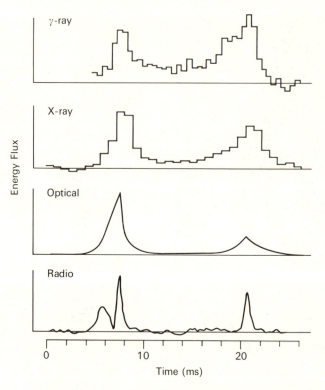

4-6 Integrated profile shapes of the Crab pulsar from radio (410 MHz) to γ-ray frequencies. The radio-frequency pulse components are, from the left, the precursor, main pulse, and interpulse. Only the main pulse and interpulse are present at higher frequencies. [Data from Manchester, 1971b; Warner *et al.*, 1969; Rappaport *et al.*, 1971; and Kurfess, 1971.]

below about 700 MHz a third pulse component, known as the precursor, can be detected preceding the main pulse. There is also evidence for additional pulse components between the main pulse and interpulse at frequencies below about 200 MHz; at these frequencies the profile is severely broadened by interstellar scattering (Rankin *et al.*, 1970), making the identification of weak components difficult. The optical profile remains significantly above the zero level between the two pulse components, as may be seen in Figure 4-6, and the fraction of the total pulsed energy between the components becomes larger at X-ray and γ-ray frequencies. Ratios of energies in the three pulse components for various frequencies are given in Table 4-1. At high γ-ray energies (> 1 GeV) the ratio of interpulse to main pulse energy is rather uncertain; McBreen and co-workers (1973) give a value of one, but other authors obtain values around three.

TABLE 4-1

Ratios of pulse energy for components of the Crab pulsar

	Precursor/ main pulse	Interpulse/ main pulse	Ref.*
Radio (410 MHz)	0.82	0.65	1
Radio (1664 MHz)	<0.03	0.20	1*
Optical	—	0.59	2
X-ray (1.5–10 keV)	—	1.10	3
X-ray (30–200 keV)	—	1.35	4
Low-energy γ-ray (100–400 keV)	—	2.3	5
γ-ray (0.6–9 MeV)	—	∼2	6,7

* REFERENCES: **1.** Manchester, 1971b.   **2.** Warner *et al.*, 1969.   **3.** Rappaport *et al.*, 1971.   **4.** Zimmermann, 1974.   **5.** Kurfess, 1971.   **6.** McBreen *et al.*, 1973. **7.** Hillier *et al.*, 1970.

Average flux densities of the pulsed emission in the range from low radio frequencies to high-energy γ-rays are shown in Figure 4-2. It is clear from these data and Table 4-1 that the radio and optical spectra are not continuous; this, together with the rather different pulse shapes and the brightness temperature and pulse fluctuation data discussed below, strongly suggests that the radio and optical emission mechanisms are different. Scattering by irregularities in the interstellar medium (discussed in Chapter 7) results in a sharp cutoff in the pulsed flux at frequencies below 100 MHz. Below this frequency the pulsar was first observed as a compact source situated within the nebula and having a very steep spectrum (Hewish and Okoye, 1964; and see Figure 4-2). The continuity of the spectra for the compact source and the pulsar, together with interferometric observations showing that the compact source is coincident with the pulsar (Vandenberg *et al.*, 1973), leave no doubt that this source is simply scattered radiation from the pulsar. Below about 100 MHz the spectral index is close to -2.0 and above this frequency it is -3.5. These values refer to the whole pulse; as described above, the spectral index differs considerably among the components.

At infrared frequencies the pulse shape is very similar to the optical profile (Becklin *et al.*, 1973) and the spectral index is about $+0.3$. The spectrum reaches a local maximum at optical frequencies and is apparently continuous with the low-energy X-ray spectrum. At about 20 keV (10^{19} Hz) the spectrum steepens, and above this energy remains essentially straight ($\alpha \approx -1.2$) to at least 1 GeV. At pulse energies less than a few keV there may be some attenuation resulting from scattering by interstellar dust (Thomas, 1975). It is possible that some or even all of the spectral curvature near 20 keV is a result of this attenuation. It is interesting that the flux from the pulsar exceeds that from the nebula at the low end of the spectrum and probably also at the high end. Assuming that the pulsar beams radiation into a solid angle of at least 0.5 steradians—the beam solid angle could be larger than this, especially at X-ray frequencies—the pulsar luminosity is not less than 3×10^{30} erg s^{-1} at radio frequencies and 7.5×10^{34} erg s^{-1} at infrared and higher frequencies. Over 90 percent of the radio luminosity is radiated at frequencies between 10 and 100 MHz; if the spectrum continues to increase steeply at frequencies below 10 MHz, the radio luminosity is substantially larger.

At both radio and optical frequencies the main pulse component of the integrated profile has a very sharp peak. At 430 MHz the width of the main pulse is less than 300 μs (Rankin *et al.*, 1970), and at optical frequencies the peak of the main pulse is unresolved with a sampling interval of 32 μs (Papaliolios *et al.*, 1970). Pulse arrival-time measurements show that the peaks of the radio and optical main pulses are emitted simultaneously to within an accuracy of 200 μs; moreover, the X-ray main pulse is coincident with the optical main pulse within 500 μs (Kurfess and Share, 1973). The existence of sharp features implies that the extent of the region emitting these features is small, less than 10 km for $\Delta t < 30$ μs. This small size in turn implies brightness temperatures of the order of 10^{29} K or more at radio frequencies. These values leave no doubt that the radio emission mechanism involves coherent radiation by many particles; at optical and X-ray frequencies, however, the brightness temperature is much lower ($\lesssim 10^{11}$ K), and coherent processes are not required.

From observations made over a three-year period at Arecibo Observatory, Rankin, Payne, and Campbell (1974) found that at 430 MHz the pulsar intensity varies by more than a factor of three, with a characteristic fluctuation time of about 30 days. The intensities of the three radio-frequency components as a function of time are shown separately in Figure 4-7. There is clearly a high correlation between the variations of the three components; Rankin, Payne, and Campbell obtained maximum correlation coefficients of about 0.96 at zero lag. At lower frequencies the intensity variations are of similar magnitude, but are not well correlated with the variations at 430 MHz, thereby implying long-term variations in the spectral index. At 196 MHz the characteristic fluctuation time is longer, about 77 days, while at 111 and 74 MHz the intensity has declined steadily over the three years. More recent observations at 408 MHz by Lyne and Thorne (1975) show that the pulsar intensity increased by about a factor of four between November 1974 and January 1975, bringing it close to the levels observed in early 1973.

At optical wavelengths the pulse shape and intensity are stable over long intervals. Horowitz, Papaliolios, and Carleton (1972) found no evidence for variations in pulse shape over a three-year period, and Groth (1975a)

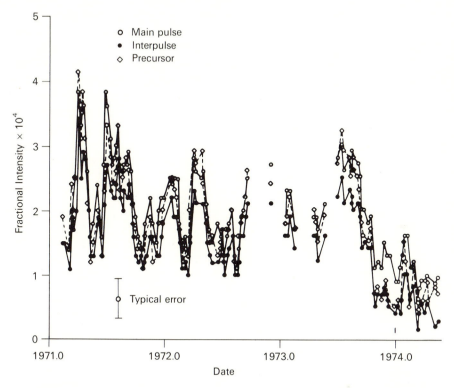

4-7 Intensities of the three components of the Crab pulsar profile at 430 MHz relative to the nebular continuum over a three-year period. [From Rankin, Payne, and Campbell, 1974.]

found that the average shape over five years, 1969 to 1974, changed by less than one percent of the peak intensity of the main pulse. All searches for evidence of long-term variations in intensity have so far proved negative.

If the optical and X-ray pulses are generated by the same emission mechanism (as suggested by the spectral continuity and similarity of pulse shapes), then the increase in size of the interpulse at X-ray frequencies implies that a small but detectable change in shape should exist across the optical band. Muncaster and Cocke (1972) found that the leading edge of the interpulse was about one percent brighter in the U band than in the V band, an effect of about the expected magnitude. Groth (1975b) compared observations made in the V and I (~ 8000 Å) bands and found that the leading edge of the interpulse is brighter in the V band by between one and two percent, also consistent with extrapolation to the X-ray profiles. For the main pulse the frequency dependence appears to be opposite, with the leading edge about one percent brighter in the I band. Becklin and co-workers (1973) found that the infrared profile at 2.2 μm differed from the optical profile by less than four percent of the main-pulse peak.

Substantial linear polarization has been detected in the integrated pulse profile at both radio and optical frequencies. The radio precursor is essentially 100 percent polarized with approximately constant position angle, but the main pulse and interpulse are only about 10 percent polarized. The position angle in these components is difficult to determine, but is probably almost constant throughout the components and about equal to the position angle of the precursor. At optical wavelengths the polarization of the pulsed emission is quite different (Figure 4-8). The fractional

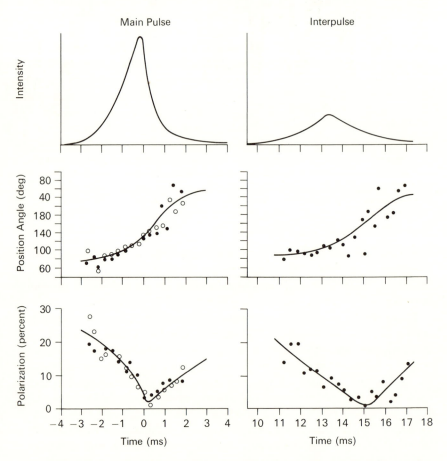

4-8 Variation of the fractional linear polarization and position angle of the electric vector for the Crab pulsar main pulse and interpulse at optical wavelengths. [From Kristian *et al.*, 1970.]

polarization has a maximum value of about 20 percent near the leading edge of the main pulse and reaches a minimum of a few percent about 300 μs after the pulse peak. The position angle variation is an S-shaped curve similar to that seen in several other pulsars at radio wavelengths, with a total swing of about 160°. The variation of polarization across the interpulse is quite similar to that of the main pulse.

After compensation for the effects of interstellar Faraday rotation, the position angle of the radio pulses is 145° ± 20° (Manchester, 1972), close to the value at the peak of the optical main pulse. Ferguson, Cocke, and Gehrels (1974) suggest that, after removal of the effects of interstellar polarization, the actual variation of position angle near the polarization minimum of the optical pulse may be more complicated. These authors also find some evidence for long-term variability in the polarization. Cocke, Muncaster, and Gehrels (1971) set a limit of 0.07 percent on the circular polarization at the peak of the main optical pulse.

Another feature of the emission from the Crab pulsar is the occasional occurrence of very strong radio pulses. In fact, it was these strong pulses that were discovered by Staelin and Reifenstein in 1968. A plot of the number of pulses observed versus flux density at a frequency of 146 MHz is shown in Figure 4-9. These data show, for example, that a pulse more

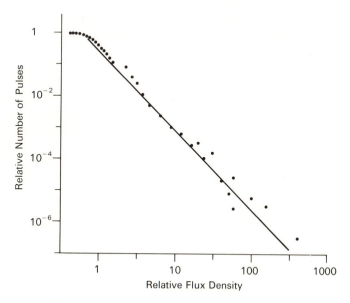

4-9 Pulse–height distribution for the Crab pulsar at 146 MHz. Flux densities are relative to the mean value. [After Arygle and Gower, 1972.]

than 10 times as strong as the average occurs about every 10^3 pulses. The straight line in Figure 4-9 represents a power law with exponent -2.5. At 430 MHz, Heiles, Campbell, and Rankin (1970) found occasional pulses with intensity greater than 1,000 times the average value. Pulse widths are typically 100 μs or less, implying brightness temperatures in excess of 10^{31} K. Despite these large intensities, the strong pulses are sufficiently infrequent that they contribute only a few percent of the pulsed radio energy. Almost all of the strong pulses occur at the longitude of the main pulse; Gower and Argyle (1972) found that at 146 MHz about seven percent occur at the longitude of the interpulse and none are detected from the precursor. In simultaneous observations at 111 and 318 MHz, Heiles and Rankin (1971) found that when a pulse is strong at one of these frequencies it generally is not strong at the other, showing that the spectrum of these pulses is variable. Giant pulses are not observed at frequencies above about 500 MHz and are in general only weakly polarized.

In contrast to the behavior at radio frequencies, there is no evidence for any short-term variation in shape or intensity of the optical pulses. Hegyi, Novick, and Thaddeus (1971) showed that for the main pulse the ratio of the mean squared intensity $\langle I^2 \rangle$ to the square of the mean intensity $\langle I \rangle^2$ is less than about 1.02. If, for example, one in 10^3 pulses is more intense than average, by implication the enhancement factor is less than five. These authors also showed that there are no statistically significant fluctuations on time scales from 1 ns to 30 μs. Short-term variability in the X-ray pulses with a time scale of a few tenths of a second may have been detected by Forman and co-workers (1974).

Among the most important observations of the Crab pulsar are the pulse arrival-time measurements, from which the pulsar period, period derivative, and other parameters are calculated. These results are described in Chapter 6, and their interpretation is discussed in Chapter 9.

5

X-Ray Pulsars
and Binary Systems

Most of the information on pulsars in this book is based on observations made at radio frequencies in the range 10 MHz to 10 GHz. There is clear evidence that at frequencies above this range pulsar flux densities generally decrease rapidly. Nevertheless, a related class of objects, which do not radiate significantly at radio frequencies, has been discovered at X-ray frequencies. The X-ray emission from these objects, known as X-ray pulsars, pulsates with observed periods in the range 0.7 to 835 s. All known X-ray pulsars are believed to be members of binary systems. By contrast, only one of the 149 known radio pulsars is clearly a member of a binary system, namely, PSR 1913 + 16, which has the second-shortest period known (59 ms) and the very short orbital period of 7.75 hours.

Observations of binary stars have played a very important role in optical astronomy because they provide the only direct method of determining stellar masses. The same advantage is afforded by the binary X-ray sources and the one known binary (radio) pulsar. Furthermore, the binary pulsar, which may be thought of as an accurate clock moving in an eccentric orbit at a speed $v/c \approx 10^{-3}$ in a strong and varying gravitational field, provides an ideal laboratory for testing gravitational theories. We shall begin this chapter by discussing what is known of the emission of ordinary pulsars outside the radio-frequency range; we shall then describe the X-ray pulsars and the binary radio pulsar and briefly explore their possible evolutionary histories.

Pulsar Emission Above the Radio-Frequency Range

Pulsars generally have rather steep radio frequency spectra (p.00), and extrapolation of these spectra to optical or higher frequencies would predict flux densities far below detectable limits. Yet several pulsars are detectable well above the radio frequency range, including the Crab pulsar (PSR 0531 + 21) and the Vela pulsar (PSR 0833 − 45). Evidently, another emission mechanism (or mechanisms) is responsible for the high frequency radiation from these pulsars. The emissions of the Crab pulsar and its associated supernova remnant were discussed in detail in the previous chapter. Like the Crab pulsar, the Vela pulsar is also an unusually rapidly pulsing source, and because it is only one-quarter the distance of the Crab pulsar it has seemed an especially likely candidate in searches for higher frequency pulsed emission.

Many searches for the Vela pulsar have been made at optical wavelengths. Until recently, all of these have been negative, in part owing to uncertainty in the radio position. In a very deep photograph of a field that did include the radio position currently believed to be the most reliable (Goss *et al.*, 1977) Lasker (1976) detected a faint blue object ($m_B \approx 23.7$) at a position within the uncertainties of the radio position. Lasker searched unsuccessfully for pulsations in the light from this object. A more recent search by Wallace and co-workers (1977) was successful in detecting pulsed optical emission from a field that included Lasker's candidate star. As shown in Figure 5-1, the pulse profile consists of two broad components separated by about 22 ms or 90° of longitude.

Measurements of relative pulse arrival times show that, after correction for dispersive delay, the radio pulse is emitted about 20 ms before the first optical pulse. The time-averaged blue magnitude of the pulsed emission is about 25, a factor of nearly 2,000 fainter than the Crab optical pulses. After taking into account the different distances and extinctions for the two pulsars, the ratio of intrinsic luminosities is about 10^5. If Lasker's blue star is in fact the pulsar, it appears that only part of its optical emission is pulsed.

X-rays in the range 1–10 keV have been detected from a point source near the Vela pulsar, and Harnden and Gorenstein (1973) reported that, near 1 keV, a fraction of the X-ray flux from this source is modulated at the radio pulsation period of the pulsar. However, subsequent observations have failed to confirm the periodic modulation, despite improved

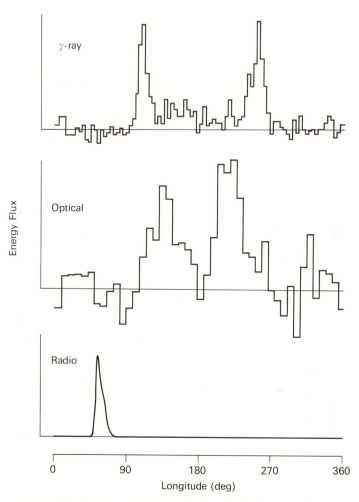

5-1 Integrated profile shapes for the Vela pulsar, PSR 0833−45, at radio (2295 MHz), optical, and γ-ray ($\gtrsim$ 50 MeV) frequencies. The pulses are aligned in relative phase as they are emitted from the pulsar. [Data from Manchester, Hamilton, *et al.*, 1976; Wallace *et al.*, 1977; and Buccheri, 1976.]

sensitivity (Moore *et al.*, 1974). Searches for higher-energy pulsed X-rays have also been unsuccessful (e.g., Rappaport *et al.*, 1974; Pravdo *et al.*, 1976).

Despite the failure to detect X-rays from the Vela pulsar, pulsed emission at γ-ray energies has been detected. Using instruments aboard the second Small Astronomy Satellite (SAS-2), Thompson and co-workers (1975) detected a broad two-component pulse. Subsequent observations by Buccheri (1976) and co-workers using the European COS-B satellite showed that the γ-ray pulse profile (Figure 5-1) was very similar to that of the Crab pulsar with the two components separated by 38 ms or 150° of longitude. As shown in Figure 5-1, the phase of the γ-ray pulses is such that the optical pulses lie symmetrically between the more widely spaced γ-ray pulses. The radio pulse is emitted 30 ms (120°) before the mid-phase point of both the optical and the γ-ray pulses. The observed pulsed flux is about 10^{-5} photons $cm^{-2} s^{-1}$, which, assuming a characteristic photon energy of 100 MeV, corresponds to a pulsed luminosity of 3×10^{33} erg s^{-1}. This value is comparable to the pulsed luminosity of the Crab pulsar in the same energy range, and is about a factor of 10^5 greater than the radio luminosity of the Vela pulsar. In earlier observations Albats and co-workers (1974) detected a narrow γ-ray pulse coincident with the radio pulse at energies between 10 and 30 MeV. These findings together with the discrepant X-ray results suggest that high-frequency pulsed radiation from the Vela pulsar may be complex and/or time-variable.

Data from the SAS-2 γ-ray experiment has been searched for emission from other radio pulsars (Ögelman *et al.*, 1976). Except for the Crab and Vela pulsars, none of the 134 objects searched had a significant time-averaged flux. Of the 134, periods sufficiently accurate to permit folding of the γ-ray data were available for 75 pulsars. Significant pulsed emission was detected for two pulsars, PSR 1747−46 and PSR 1818−04. For PSR 1818−04 the phase of the γ-ray pulse was not the same as the radio-pulse phase; for PSR 1747−46 the radio phase was not known. These two pulsars have relatively long periods (0.742 s and 0.598 s, respectively) and generally seem rather undistinguished. In both cases the implied γ-ray luminosity (about 10^{33} and 10^{34} erg s^{-1}, respectively) represents a large fraction of the total rate of energy loss from the rotating neutron star.

Searches for other known radio-frequency pulsars at infrared and optical wavelengths have all been fruitless, as have been searches of galactic and extragalactic supernova sites. The objects studied include many known radio pulsars, especially those of short period, as well as white dwarfs, planetary nebulae, novae, and supernovae.*

Binary X-Ray Sources

A number of X-ray sources discovered using rocket- or satellite-borne instruments, e.g., the *Uhuru* orbiting observatory, have been optically

---

* Summaries of searches at optical and infrared wavelengths have been published. See for example, Kristian, 1970b; Horowitz *et al.*, 1971; and Papaliolios and Horowitz, 1973.

identified with close binary systems. Some interesting parameters of seven of these systems are listed in Table 5-1. In each of the seven, the binary system is believed to consist of a collapsed object, probably a neutron star or a black hole, and a main sequence or post-main sequence star that is normal except for tidal distortion and asymmetrical heating caused by the nearby X-ray source. The binary nature of these systems is revealed by periodic variations in both the optical and X-ray fluxes.

Optical variations (Δm_v in Table 5-1) usually amount to about 0.1 magnitude and are believed to result from some combination of tidal distortion of the optical star and X-ray heating of one face of the star. X-ray eclipses are observed in five of the seven systems when the compact star passes behind the main-sequence companion. The orbital periods, typically a few days in length, are also manifested spectroscopically by periodic variations in the radial velocity of the visible star. If the velocity curve can be shown to reflect the motion of the visible star about the center of mass of the system, i.e., to be free of complicating effects such as gas streaming, then it yields all of the information available from any normal single-line spectroscopic binary star, including orbital period, eccentricity, length of the projected semi-major axis, and the so-called mass function, defined by

$$f_1(M_1, M_2, i) = \frac{(M_2 \sin i)^3}{(M_1 + M_2)^2} = \frac{4\pi^2}{G} \frac{(a_1 \sin i)^3}{P_b^2}. \tag{5-1}$$

In this expression, which is a straightforward result of Newton's laws for elliptical orbits, M_1 is the mass of the star whose radial velocity has been measured; M_2 is the mass of the companion; i is the inclination between the plane of the orbit and the plane of the sky; G is the gravitational constant; $a_1 \sin i$ is the projected semi-major axis of the orbit of star 1, and P_b is the period of the binary orbit. Although evaluation of this expression is not in itself sufficient to determine the two masses, it places a useful constraint on the relationships between M_1, M_2, and i.

The X-ray flux of four of the sources in Table 5-1 is also periodically modulated, and three of these sources have periods comparable to those of radio pulsars. Integrated profiles of these three "X-ray pulsars" (Hercules X-1, Centaurus X-3, and SMC X-1) are shown in Figure 5-2. In contrast to radio pulsars, the integrated profiles of X-ray pulsars seem to vary considerably in shape over time scales of days and weeks (Tuohy, 1976; Lucke *et al.*, 1976). For example, in *Uhuru* observations (Giacconi, 1975a) the integrated profile of Centaurus X-3 was single-peaked, rather than double-peaked as it is in Figure 5-2. Despite such shape variations, the modulation periods of the X-ray sources are sufficiently stable to obtain velocity curves for the sources. This allows

TABLE 5-1

Parameters of X-ray binary systems

X-ray source *	α(1950)	*l*	Distance (kpc)	Orbital period (days)
Companion star	δ(1950)	*b*		
SMC X-1 (3U 0115−73)	01ʰ15ᵐ44ˢ3	300°	65 ± 10	3.8927 ± 10
SK 160	−73°42′ 54″	−44°		
Vela X-1 (3U 0900−40)	09 00 13.2	263	1.4 ± 0.3	8.95 ± 2
HD 77581	−40 21 25	+4		
Centaurus X-3 (3U 1118−60)	11 19 03	292	5−10	2.087129 ± 7
Krzeminski's star	−60 21 00	+0.4		
Scorpio X-1 (3U 1617−15)	16 17 04.3	359	0.3−1	0.787313 ± 1
V 818 Sco	−15 31 13	+24		
Hercules X-1 (3U 1653+35)	16 56 01.7	58	2−6	1.700165 ± 2
HZ Her	+35 25 05	+38		
3U 1700-37	17 00 32.7	348	1.5 ± 0.5	3.4120 ± 3
HD 153919	−37 46 27	+2		
Cygnus X-1 (3U 1956+35)	19 56 28.8	71	2.5 ± 0.5	5.5999 ± 9
HDE 226868	+35 03 55	+3		

* The 3U names refer to the Third *Uhuru* catalog of X-ray sources (Giacconi *et al.*, 1974).

Visible star			X-ray source		
Spectral type	$m_v(\Delta m_v)$	Mass $(M_\odot)$	Mass $(M_\odot)$	Eclipse duration (orbital periods)	Short-term variability
B0.5I	13.3 (0.09)	26–30	~2.2–4.2	0.14	Periodic at 0.7157 s
B0.5Ib	6.9 (0.07)	18.5–24	1.35–1.9	0.19	Nonperiodic down to 1 s; periodic at 283 s
O6.5II	13.4 (0.08)	16.5–20	0.6–1.8	0.25	Periodic at 4.842 s
No stellar absorption lines	~13 (0.2)	<2	~1	None observed	Nonperiodic down to 1 s
Late A or early F	~14 (~1)	~2	~1	0.16	Periodic at 1.23782 s
O6f	6.6 (0.04)	>10	≥0.6	0.32	Nonperiodic down to 0.1 s
O9.7Iab	8.9 —	>10	9–15	None observed	Quasiperiodic down to 1 ms

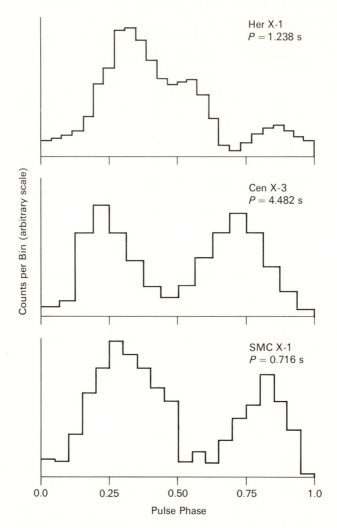

5-2 Integrated pulse profiles for "X-ray pulsars" Her X-1 (2–6 keV), Cen X-3 (3–9 keV), and SMC X-1 (1.6–10 keV). [Data from Joss and Fechner, 1975; Tuohy, 1976; and Lucke *et al.*, 1976.]

separate determination of the masses of the optical star, M_1, and the X-ray source, M_2. As shown in Table 5-1, the visible star is in most cases rather massive ($2-30\ M_\odot$), whereas the mass of the X-ray source is less ($1-3\ M_\odot$). In addition to the short-period modulation, the X-ray emission from Hercules X-1 is modulated with a 35-day period, being "on" for about 10 days and "off" for 25.

Evidence suggests that the X-radiation from X-ray binaries is produced by accretion of matter from the optical primary star onto a compact secondary. The rapid time variations observed (see last column of Table 5-1) indicate a compact source, and accretion is suggested by the very high temperatures ($\sim 10^8$ K) deduced from X-ray spectral measurements and the large X-ray luminosities (10^{36} to 10^{38} erg s^{-1}). The only type of compact star compatible with the observed range of mass for the periodic X-ray sources is a neutron star (although it is possible that the X-ray source Cygnus X-1 is a black hole).

As for the radio pulsars, the most likely origin for the periodic modulation is rotation of the neutron star. Regular pulsations in the X-ray intensity could arise if the accreting material is channeled into polar regions by a strong magnetic field inclined to the rotation axis. This model suggests that X-ray pulsars are physically very similar to radio pulsars except that as members of binary systems they are subject to accretion of material from their companion stars. Searches for pulsed radio emission from these objects have all been unsuccessful. It is probable that the accreting gas is sufficiently dense to prevent escape of radio emission from the pulsar (Illarionov and Sunyaev, 1975).

More direct evidence for accretion is provided by spectroscopic data that indicate gas outflow from the visible star (e.g., Smith *et al.*, 1973; Hensberge *et al.*, 1973), and by the observed speeding-up, on the average, of the pulsation rates of the X-ray pulsars Hercules X-1 and Centaurus X-3. These period changes are variable (Figure 5-3), but have a trend in the opposite sense to that expected from an emission process in which

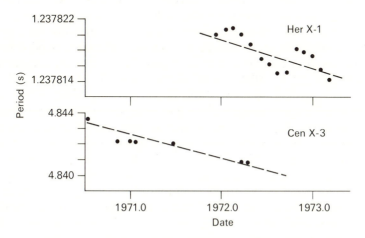

5-3 Changes in the periods of X-ray pulsations from Her X-1 and Cen X-3. [After Giacconi, 1975b.]

the radiated energy is provided by the loss of rotational energy (as for radio pulsars). An accretion process can transfer angular momentum to the neutron star in such a way as to speed up the rotation. Computed rates of period decrease are consistent with the mean rates observed.

The X-ray luminosity is approximately equal to the rate at which the accreting material loses potential energy, that is,

$$L_x \approx \frac{GM\dot{M}}{R},$$ (5-2)

where M and R are respectively the mass and radius of the compact star and $\dot{M}$ is the accretion rate. Material from the primary star could be transferred by one of two processes: (1) expansion of the star (as it evolves off the main sequence), which would cause it to overflow its Roche lobe and transfer mass through the inner Lagrangian point L_1; or (2) a well-developed stellar wind, driven by the natural luminosity of the primary star or by excess heating of its outer layers by the X-ray source. Both possibilities are diagrammatically illustrated in Figure 5-4.

Equation 5-2 suggests that the observed luminosities can be produced by accretion rates in the range 10^{-10} to 10^{-8} $M_\odot$ per year. In fact, the upper limit of the observed luminosities, 10^{38} erg s^{-1}, is close to the self-limiting point or "Eddington luminosity" at which gravitational forces are balanced by radiation pressure. Consequently, the luminosity remains close to 10^{38} erg s^{-1} even for larger mass transfer rates, up to about 10^{-6} $M_\odot$ per year, at which point the X-ray source is expected to be extinguished by absorption (van den Heuvel, 1975a). For evolved stars with fully developed Roche-lobe overflow, the mass transfer rate would exceed 10^{-6} $M_\odot$ per year when the stellar mass exceeds about 2 $M_\odot$. On the other hand, stellar winds sufficient to excite an X-ray source will not be produced by stars with mass $\lesssim 15$ $M_\odot$. Thus, the implied rates of accretion are consistent with those expected from evolving companion stars of mass $\lesssim 2$ $M_\odot$ (Roche-lobe overflow) or $\gtrsim 15$ $M_\odot$ (stellar wind).

The short orbital periods of the binary systems indicate that the orbital semi-major axes are only a few times larger than the primary stars themselves. Thus, it is not surprising that periodic eclipses of the X-ray sources are often observed. Such an eclipse is shown in the X-ray data for the source 3U 1700—37 in Figure 5-5. In this example the duration of the X-ray eclipse at orbit phase 0.0 $\pm$ 0.16 shows that the inclination i must be reasonably close to 90°, and that the occulting star must have a radius at least 0.85 times the orbital semimajor axis. If the orbital inclination is close to 90°, it is probable that the rotation axis of the neutron star is also approximately perpendicular to the line of sight. Since the X-radiation is beamed from the magnetic polar regions, the

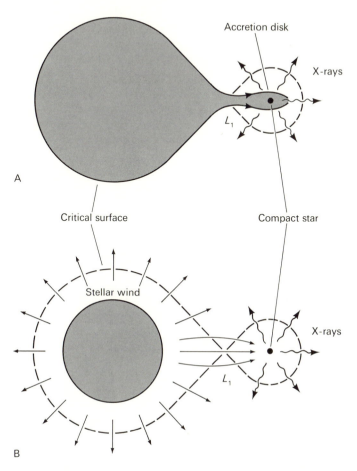

5-4 Models of binary X-ray sources powered by accretion due to (A) Roche-lobe overflow, and (B) a stellar wind. L_1 is the inner Lagrangian point.

magnetic axis must then be approximately perpendicular to the rotation axis. This is consistent with the fact that the integrated profiles of the X-ray pulsars (Figure 5-2) often have two peaks per period spaced by about half a period, i.e., one from each pole of a basically dipole field.

Data illustrating variations in the optical magnitude and radial velocity of the visible star HD 153919 (companion of 3U 1700 − 37) are also shown in Figure 5-5; the relative positions of the visible star and the X-ray source are indicated schematically at the top of the figure. In general, the radial velocity curves for binary X-ray sources are sinusoidal, or nearly so, indicating orbits of low eccentricity. On the other hand, the

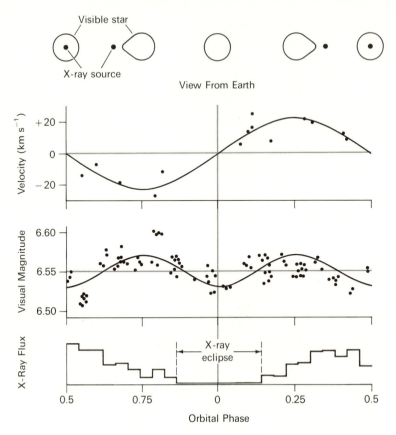

5-5 Observations of the binary system comprised of the visible star HD 153919 and the X-ray source 3U 1700−37: radial velocity (Wolff and Morrison, 1974) and visual magnitude (Jones and Liller, 1973) of the visible star, and the 6–10 keV X-ray flux (Jones et al., 1973) plotted as a function of orbital phase. These data were recorded over an extended period and folded modulo the 3.412 day orbital period. A model of the orbiting X-ray source and its tidally distorted companion star is shown at the top.

small-amplitude variations in light are more complicated, often showing two maxima and two minima per orbital period. The variations can be interpreted as due to the changing aspect of the tidally distorted and asymmetrically heated primary star. Careful analysis of the "ellipsoidal variations" exhibited in the light curve, together with data on the duration of X-ray eclipse and evaluation of one of the mass functions, can yield a self-consistent solution for both masses and the size of the larger star (see for example Avni and Bahcall, 1975). Additional observations, such as determination of the second mass function, can give independent verification of the results. Such lines of argument show that the visible stars of the more luminous binary X-ray sources (including all of those listed in Table 5-1) fill, or nearly fill, their critical Roche lobes. In some of the less luminous sources, such as 3U 0532 + 30 (or X Per), the size of the visible star is apparently much less than the critical size. This would suggest a smaller accretion rate, which accords well with the lower X-ray luminosity observed.

In addition to the three "fast" X-ray pulsars, six other X-ray sources are known to have periodic intensity variations over time scales of 100 to 835 s. These include A 0535 + 26 ($P = 104$ s), GX 1 + 4 ($P = 122$ s), A 1118 − 61 ($P = 405$ s), GX 301 − 2 ($P = 696$ s), and 3U 0532 + 30 ($P = 835$ s), as well as Vela X-1 listed in Table 5-1.* Not all of these sources have been shown conclusively to be binary objects. However, an orbiting companion is probable in each case. One of the slowly pulsing X-ray sources, Vela X-1, is an eclipsing X-ray binary with an orbit period of 8.95 days and a pulse period of 283.795 s (Clark, 1975). Evaluation of the two mass functions for this source has shown that the optical primary has a mass of $21.2 \pm 2.6\ M_\odot$ and the X-ray secondary a mass of $1.61 \pm 0.27\ M_\odot$ (van Paradijs *et al.*, 1976). This latter mass is above the upper mass limit for white dwarf stars.

Much work has been done on the possible evolutionary histories of binary X-ray sources, but before discussing this work we shall introduce the subject of binary radio pulsars and describe the only such system known at present.

The Possibility of Binary Pulsars

Since approximately half of the stars in the Milky Way are believed to be members of binary or more complicated orbiting systems, it is noteworthy that radio pulsars are, in general, solitary objects. The relatively massive stars thought to be the progenitors of supernovae, and hence of neutron stars, are especially common in binary systems. Calculations suggest that a substantial fraction of neutron stars created in binary systems will remain orbitally bound to their unexploded companions (see for example Wheeler *et al.*, 1975), therefore the very low incidence of pulsars in orbiting systems is rather surprising. On the other hand, we know of no physical requirement that rules out the possibility of radio pulsars in binary systems.

---

* Names with an A prefix refer to X-ray sources discovered by the *Ariel V* satellite (Villa *et al.*, 1976).

The information potentially available from such systems is very great. Timing measurements for radio pulsars (Chapter 6) are several orders of magnitude more precise than those currently achievable for X-ray pulsars; consequently, it might be possible to determine the orbit of a hypothetical orbiting pulsar sufficiently well to specify the masses of both components, independent of any astrophysical assumptions. Moreover, it might even be possible to "overdetermine" the orbit, so that timing observations would provide a test of gravitation theories (Hoffman, 1968).

As discussed in Chapter 6, pulse-timing stabilities are typically such that arrival-time residuals are less than 1 ms over time scales of several years. Therefore, if a pulsar existed in an orbit that caused it to move about the system barycenter with an amplitude exceeding ～300 km, the motion would be detectable as a periodic term in the residuals. Even orbiting companions as small as Earth could be detected in this way, for the annual motion of Earth displaces the sun from the Earth–Sun center of mass by approximately 0.0015 light-seconds. The corresponding amplitude for the Jupiter–Sun system is 2.6 light-seconds.

The constraints placed on pulsar binary motions by existing observations have been considered quantitatively by Lamb and Lamb (1976). The principal conclusions of this work are summarized in Figure 5-6,

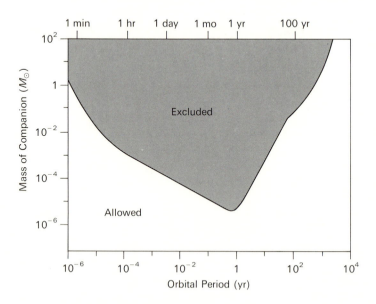

5-6 Pulsars that have been carefully timed over intervals of a year or more and which show timing residuals <1 ms cannot have orbiting companions whose mass and period fall in the shaded region. [After Lamb and Lamb, 1976.]

which shows the combinations of companion mass and orbital period that are excluded by timing observations, provided these observations are made over intervals of a year or more and show residuals of less than 1 ms. It is clear that orbiting companions are generally ruled out, except those with very short ($\lesssim$ minutes) or very long ($\gtrsim$ decades) periods, or those with very small mass ($\lesssim 0.1\ M_\odot$ for the shortest and longest periods, and $\lesssim 10^{-4}\ M_\odot$ for periods between one day and five years). Except for PSR 1913+16, none of the 50 or so carefully observed pulsars can have orbiting companions approaching stellar mass.

Binary Pulsar PSR 1913+16

The binary pulsar PSR 1913+16 was first detected in July, 1974 during a systematic search of the galactic plane for new pulsars (Hulse and Taylor, 1975a, b). This pulsar was of immediate interest because its period of 0.059 s was less than that of any other known pulsar except the one in the Crab Nebula. It soon became clear that the large cyclic variations observed in its period could be easily understood if the pulsar was in orbit about another massive object. A velocity curve derived from observations during September 12–26, 1974 (Figure 5-7) was shown to be consistent with an orbit of projected semimajor axis $a_1 \sin i = 1.0\ R_\odot$, eccentricity $e = 0.62$, period $P_b = 7.75$ hours, and mass function $f_1(M_1, M_2, i) = 0.13\ M_\odot$. These facts and the absence of observable eclipses were sufficient to show that the unseen companion was also a compact object, with mass comparable to that of the pulsar. In due course, such effects as transverse Doppler shift, gravitational redshift, and relativistic advance of periastron should be easily measurable, which would allow the masses to be determined and a number of interesting gravitational and relativistic phenomena to be studied.

PSR 1913+16 lies at galactic coordinates $l = 49°.9$, $b = 2°.1$, and its dispersion measure of 167 cm^{-3} pc suggests a distance of about 5 kpc. Kristian, Clardy, and Westphal (1976) have shown that there is no visible object brighter than about 21st magnitude at the position of the radio pulsar. This corresponds to an absolute magnitude $M_v \gtrsim 3$ for the binary system. Synchronous averaging at the pulsar period gives a more stringent limit of $m_v > 23$ for the time-averaged visual magnitude of periodic fluctuations. X-ray observations of the region in the 2.5–7.5 keV range, utilizing a collimated proportional counter aboard the satellite OAO *Copernicus*, have yielded an upper limit of 3×10^{35} erg s^{-1} for the time-averaged X-ray luminosity (Davidsen *et al.*, 1975). Radio observations designed to detect the companion if it were also a pulsar have yielded an upper limit of 60 μJy for the time-averaged pulsed flux at 430 MHz (Taylor *et al.*, 1976). Thus, at the present time PSR 1913+16 is

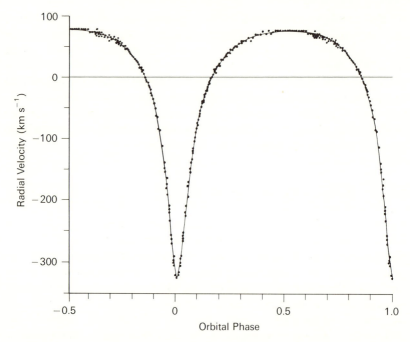

5-7 Velocity curve of the binary pulsar PSR 1913+16. [From Hulse and Taylor, 1975a.]

detectable only at radio wavelengths, and its companion has not been directly observed.

By the end of 1975, accurate timing data for PSR 1913+16 had been accumulated for more than a year. These arrival-time data were analyzed using a development of the standard least-squares treatment that gives the orbital elements as well as the pulsar period, its derivative, and the celestial coordinates (see p. 106ff.). The results are summarized in Table 5-2. A diagram of the orbit of PSR 1913+16, illustrating definitions of a_1 and ω, is shown in Figure 5-8.

The effects of second-order Doppler shift and gravitational redshift are combined in a further term,

$$\gamma = \frac{2\pi a_1^2 e}{c^2 P_b}\left(2 + \frac{M_1}{M_2}\right), \tag{5-3}$$

which in principle can be determined from the timing measurements (see p. 109). However, the coefficient of γ depends on orbital phase in almost the same way as other terms in the least-squares expansion. It is expected that after several years the longitude of periastron will have moved

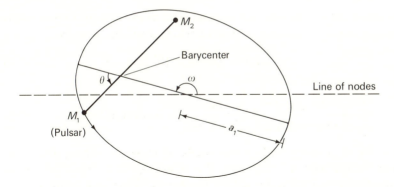

5-8 Diagram of the orbit of the binary pulsar PSR 1913+16 seen face-on. The line of nodes is the intersection of the plane of the orbit and the plane of the sky, and the inclination angle i is the angle between these two planes. (ω = longitude of periastron; a_1 = semimajor axis of the orbit).

sufficiently to decouple these terms and hence to permit evaluation of γ. It should be noted that the value of $a_1 \sin i$ quoted in Table 5-2 has "absorbed" this term and is therefore systematically large by a factor of ~ 1.002.

For a complete understanding of the dynamics of this orbiting system, one would need to determine independently the two masses M_1 and M_2,

TABLE 5-2

Parameters of PSR 1913+16 [From Taylor et al., 1976.]

Right ascension	$\alpha(1950.0) = 19^h13^m12\overset{s}{.}484 \pm 0\overset{s}{.}008$
Declination	$\delta(1950.0) = +16°01'08\overset{''}{.}4 \pm 0\overset{''}{.}2$
Period	$P = 0.059029995272 \pm 5$ s
Derivative of P	$\dot{P} = (8.8 \pm 0.3) \times 10^{-18}$ s s^{-1}
Orbital semimajor axis	$a_1 \sin i \equiv x = (7.0043 \pm 0.0004) \times 10^{10}$ cm
Orbital eccentricity	$e = 0.61717 \pm 0.00005$
Binary period	$P_b = 27906.980 \pm 0.002$ s
Longitude of periastron	$\omega_0 = 178\overset{\circ}{.}861 \pm 0\overset{\circ}{.}007$
Time of periastron passage	$T_0 = $ JD 2442321.433210 $\pm$ 0.000004
Rate of advance of periastron	$\dot{\omega} = 4.22 \pm 0.04$ deg yr^{-1}
Derivative of x	$\dot{x} = -0.2 \pm 1.2$ cm s^{-1}
Derivative of e	$\dot{e} = (1 \pm 1) \times 10^{-11}$ s^{-1}
Derivative of P_b	$\dot{P}_b = (2 \pm 6) \times 10^{-10}$ s s^{-1}

the orbital semimajor axis (a_1), and the orbit inclination (i). Two relations between the four quantities are immediately available, namely

$$a_1 \sin i = 7.0043 \times 10^{10} \text{ cm} \tag{5-4}$$

and

$$f_1 = \frac{(M_2 \sin i)^3}{(M_1 + M_2)^2} = \frac{4\pi^2}{G} \frac{(a_1 \sin i)^3}{P_b^2} = 0.13126 \, M_\odot. \tag{5-5}$$

In order to proceed further, one needs to know something about the nature of the unseen companion. If it is sufficiently compact to behave dynamically as a point mass, then the general relativistic prediction for $\dot{\omega}$ provides a third relation between the four unknown quantities (see for example Landau and Lifshitz, 1962):

$$\dot{\omega} = \left[\frac{6\pi G}{c^2 P_b a_1 \sin i(1 - e^2)} \right] [M_2 \sin i]. \tag{5-6}$$

The expression has been written in this form to emphasize that the first term in brackets is already determined. From Equations 5-5 and 5-6, and inserting numerical values for the known quantities, we obtain an expression for the total system mass:

$$\mathcal{M} \equiv \frac{M_1 + M_2}{M_\odot} = \left(\frac{\dot{\omega}}{2.11 \text{ deg yr}^{-1}} \right)^{3/2}. \tag{5-7}$$

With the observed value of $\dot{\omega} = 4.22$ deg yr^{-1}, we find $\mathcal{M} = 2.83$. Combining this value for the total mass with the value of the mass function (Eqn. 5-5) shows that the orbit inclination must satisfy $21° < i \leq 90°$. The distribution of the total mass between M_1 and M_2 is then a simple function of i, as illustrated in Figure 5-9.

A fourth relationship giving an unambiguous determination of the four quantities would be provided by measurement of the second-order Doppler and gravitational redshift term, γ (Eqn. 5-3). With the aid of known values of $a_1 \sin i$, e, P_b, and $\dot{\omega}$, and using Equations 5-5 and 5-7, one can rewrite Equation 5-3 in terms of M_2 alone (Blandford and Teukolsky, 1976):

$$\gamma = 2.07 \times 10^{-3} \left(\frac{M_2}{M_\odot} \right) \left(1 + \frac{M_2}{2.83 \, M_\odot} \right) \quad \text{(sec)}. \tag{5-8}$$

The restrictions already placed on M_2 (Figure 5-9) require γ to be in the range 0.003 s $< \gamma < 0.012$ s.

Is the unseen companion of PSR 1913+16 sufficiently compact to ensure that tidally or spin-induced distortions of its shape will not cause

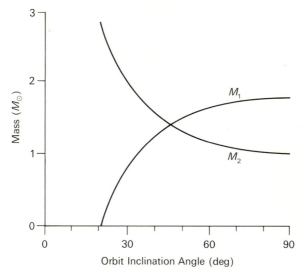

5-9 Possible masses of PSR 1913 + 16 and its orbiting companion (M_1 and M_2, respectively), plotted as a function of orbit inclination angle, i. The curves are valid if the only significant contribution to $\dot{\omega}$ is the general relativistic one. [From Taylor *et al.*, 1976.]

significant classical contributions to $\dot{\omega}$? This question has not yet been answered. However, there exists only a small range of stellar objects that are not already ruled out by the observed absence of eclipses and rate of advance of periastron, and yet are too large to behave as point masses (Roberts *et al.*, 1976). Smarr and Blandford (1976) have shown that only a helium main-sequence star and a rapidly rotating white dwarf are candidates for this "in-between" category. Moreover, Webbink (1975) has argued on evolutionary grounds that a neutron star is the most likely companion. If this can be shown to be the case, then timing measurements of PSR 1913 + 16 should provide some previously unavailable tests of gravitation theories.

Several such tests that appear to be both observationally feasible and theoretically illuminating have been suggested. Most theories of gravity predict a secular decrease in the binary period due to a loss of energy in the form of gravitational waves. For example, general relativity predicts a value (Wagoner, 1975)

$$\dot{P}_b = -0.85 \times 10^{-12} \left(\frac{\mathcal{M}^{4/3}}{\sin i}\right)\left(1 - \frac{0.51}{\mathcal{M}^{1/3} \sin i}\right) \approx -3 \times 10^{-12}. \quad (5\text{-}9)$$

For comparison, Eardley (1975) has pointed out that the Brans–Dicke theory allows dipole as well as quadrupole gravitational radiation and predicts a value of $\dot{P}_b$ as much as three orders of magnitude larger. In a few years the accuracy with which $\dot{P}_b$ and i can be determined should be sufficient to allow tests of both the Brans–Dicke and the general relativistic predictions.

According to some metric theories of gravity, the center of mass of a binary system may be accelerated toward periastron because of a violation of post-Newtonian momentum conservation (Will, 1976). Observations of secular changes of period (of either the pulsar or the orbit) can be used to place limits on such an acceleration, and hence on the parametrized post-Newtonian (PPN) parameter ζ_2. The contribution to $\dot{P}$ from this cause would be (Will, 1976)

$$\dot{P} \approx -4 \times 10^{-16} \zeta_2 \frac{X(1 - X)}{(1 + X)^2} \mathcal{M}^{2/3} T, \qquad (5\text{-}10)$$

where $X = M_1/M_2$ is the mass ratio and T is the time in years spanned by the observations. From the values $\dot{P} = 8.8 \times 10^{-18}$, $\mathcal{M} = 2.83$, and $T = 1$, we obtain

$$|\zeta_2| < 0.01 \frac{(1 + X)^2}{X(1 - X)}. \qquad (5\text{-}11)$$

Figure 5-9 shows that unless $|i - 46°|$ is less than a few degrees, we will have $|1 - X| > 0.1$, and the limit on ζ_2 would become at least as restrictive as $|\zeta_2| < 0.4$. The best limit now independently available is $|\zeta_2| < 100$, obtained from gravitational redshift data for white dwarfs.

Other possible uses of PSR 1913+16 to test gravitation theories have been discussed by Esposito and Harrison (1975), Hari-Dass and Radha-krishnan (1975), Smarr and Blandford (1976), and others. The most promising of these involve measurements of third-order terms—terms of order $P_b(v/c)^3$—in the orbital variation of pulse arrival times, and spin precession of the pulsar axis, both of which may be feasible with improved observations.

Evolution of Close Binary Systems

The problem of understanding the evolution of close binary systems is much more complicated than the corresponding problem for single stars because there is ample evidence that large quantities of mass are exchanged between many pairs of binary stars. Reviews of the subject have been published by Plavec (1968) and Paczynski (1971), and we will not attempt to treat the problem in detail here. Instead we shall, for illustrative purposes, describe a possible evolutionary scenario for compact objects in binary systems.

The scenario, depicted in Figure 5-10, is based on the model proposed by van den Heuvel (1975b). It begins with two upper main-sequence stars of mass 20 $M_\odot$ and 8 $M_\odot$, in an orbit of period 4.5 days (stage 1). After about six million years the 20-$M_\odot$ star evolves off the main sequence, expands to fill its critical Roche surface, and begins to deposit mass onto its companion (stage 2). In an interval of only about 30,000 years, nearly 15 $M_\odot$ of material is transferred onto the secondary, leaving a Wolf–Rayet or helium star of about 5 $M_\odot$ and a main sequence star of about 23 $M_\odot$ (stage 3). Conservation of angular momentum has lengthened the binary period to about 11 days at this stage. After another half-million years or so, the helium star explodes as a supernova, ejecting about 3 $M_\odot$ of gas and leaving a compact remnant of 2 $M_\odot$, presumably a neutron star (stage 4). The orbital period has by now increased to nearly 13 days. The 23-$M_\odot$ star continues to evolve, and after about another four million years it becomes a blue supergiant with a strong stellar wind (stage 5). Accretion turns the compact star into a powerful X-ray source. This stage lasts only about 40,000 years, for the supergiant soon expands to fill its Roche lobe, and a second stage of mass exchange begins in which the X-ray source is extinguished by the excessive accretion rate (stage 6). Because the compact star can accept only a small fraction of the out-flowing gas, some 17 $M_\odot$ of matter is lost from the system. At the completion of this mass loss stage, perhaps 200,000 years later, there remains a compact star of 2 $M_\odot$ and a newly formed helium star of about 6 $M_\odot$ in a close orbit of $P_b \approx 0.2$ day (stage 7). Finally, when the helium star becomes a supernova, leaving behind a second compact star, the system will consist of two compact stars in a close orbit, or, if the orbit becomes unbound, two "runaway" compact stars (stage 8).

From stage 4 onward the system contains an object potentially observable as a radio pulsar, but detectable radio emission will not escape until the orbit is sufficiently free of plasma to allow radio waves to propagate. Thus, one would not expect to observe binary pulsars with massive main sequence companions. However, by the time stage 8 is reached, the system is cleansed of all matter except that contained in the two compact stars, and one can imagine that either or both of these objects could be observable radio pulsars—whether they remain orbitally bound together or not.

The durations of the various stages of this evolutionary sequence are reasonably well determined from quantitative calculations, enabling us to apply some observational tests to the scenario just described. Van den Heuvel (1975b) does this by arguing that there are about 4,000

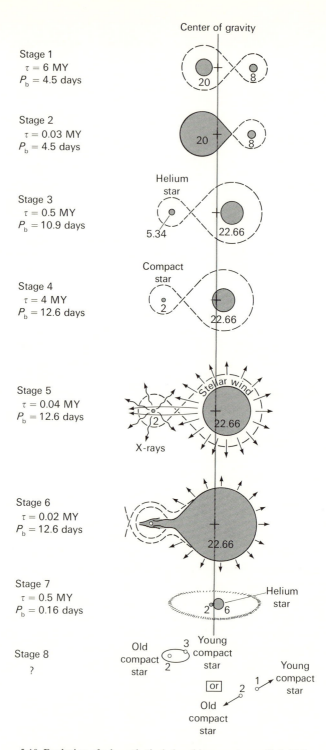

5-10 Evolution of a hypothetical close binary system with initial masses of 8 and 20 $M_\odot$. Each stage is labeled with its approximate duration (τ) in millions of years and the period of the binary orbit (P_b) in days. The numbers inside the representations of the stars indicate mass ($M_\odot$). See p. 97 for details. [After van den Heuvel, 1975b.]

TABLE 5-3

Number of objects in the Galaxy in different stages of evolution into compact binary stars. Stages are described in Figure 5-10 [After van den Heuvel, 1975b]

	Lifetime $(10^6$ yr)	Expected number in galaxy	Within 3 kpc distance	
			Expected	Observed
Stage 1: Unevolved close binary ($M_1 > 15\,M_\odot$)	6.2	4,000[a]	200[a]	200[a]
Stage 3: Wolf–Rayet binary	0.56	360	18	~17 (34 W–R stars, of which 50% are binaries)
Stage 4: Main-sequence star plus collapsed star (X-ray quiet)	3.6	2,400	120	?
Stage 5: Massive X-ray binary	0.02–0.05	13–32	0.6–1.6	3
Stage 8: Binary pulsar	2?	1,200fb[b]	60fb[b]	0.3[c]

[a] These are assumed (input) numbers.

[b] Factor f is the probability of the orbit surviving a second supernova outburst; factor b is the fraction of pulsars observable after allowing for beam effects.

[c] Actually one object at 5 kpc distance.

unevolved compact binaries in the Milky Way with one member more massive than 15 $M_\odot$. Then, assuming that such pairs evolve according to the preceding model, one should expect the number of objects in each stage of evolution to be directly proportional to the lifetime of that stage. This argument is given in numerical form in Table 5-3; one can see that the observed numbers of Wolf–Rayet binaries and massive X-ray binaries are in satisfactory agreement with the theory. Allowing for the fact that we can observe only about one-fifth of existing pulsars (because of beaming effects), the observed number of binary pulsars suggests empirically that a close binary system of total mass ~8 $M_\odot$ (stage 7 in Figure 5-10) will survive a supernova explosion of the larger member only about three percent of the time. Consequently, a population of pulsars should exist that possess rather large "runaway" velocities. This too is in agreement with observation, as will be discussed in Chapter 8.

6

Pulse Timing Observations

The most remarkable characteristic of pulsars is undoubtedly the stability of their basic pulsation periods. The earliest observations (Hewish *et al.*, 1968) showed that the period of PSR 1919+21 was constant to better than one part in 10^7, already good enough to limit severely the range of possible pulsar models. Subsequent measurements have shown that the periods are not quite constant but that in some cases they are stable and predictable to better than one part in 10^{12} over several years, almost as good as an atomic frequency standard. As described in Chapter 1, a rotating neutron star seems to be the only astrophysical system capable of generating a periodic pulse train of such stability with periods in the observed range.

In this chapter we shall describe the techniques used to measure pulsar periods accurately, as well as the results of such observations. Because pulse arrival times at the earth depend on the earth's motions in space and the celestial coordinates of the pulsar, an accurate ephemeris giving the position of the earth (with respect to the solar system barycenter) can be used to derive accurate pulsar positions. We shall also discuss the two classes of period variations observed—steady secular changes and unpredictable irregularities—and the analysis of arrival-time data for binary pulsars.

Observational Techniques

The initial determination of the period of a pulsar is normally made as part of the discovery process, either by directly counting the number of pulses in a given interval or by analyzing a data stream for periodic fluctuations. The period can then be improved by measuring pulse arrival times separated by longer and longer intervals. At each stage of this process it is necessary that the period and arrival times be sufficiently accurate that there is no ambiguity in the number of elapsed periods between observations.

Pulse arrival times are normally determined by fitting a standard pulse shape to an integrated profile. For high accuracy it is important that the profile shape be stable; as described in Chapter 2 (p. 30), summation of a few thousand pulses is generally sufficient to ensure this. Because many pulsars have a high degree of linear polarization, and because Faraday rotation in the earth's ionosphere varies significantly with time of day, it is important that either total intensity or one of the circular polarizations be recorded. By cross-correlating a standard profile with an observed integrated profile, the pulse phase can usually be determined to within 10^{-4} periods, at least for the stronger pulsars.

To convert this phase to a pulse arrival time, the epoch at some point, usually the start of the integration, must be known. For detailed studies of pulsar periods and their variations, only atomic frequency standards, used either directly or via a radio link (such as Loran-C), can give epochs to within the required accuracy of a few microseconds over many months. Radio pulse arrival times at the telescope typically have an uncertainty of about 100 μs. Arrival times for the Crab pulsar can be obtained from either the radio or the optical pulses. Uncertainties are comparable for the two types of measurement, about 10 μs from an observation of duration about one hour (Roberts and Richards, 1971; Groth, 1975a).

The effects of the earth's motion on the observed arrival times may be corrected for by referring arrival times to the solar system barycenter. An ephemeris of the earth's motion derived from planetary radar observations (see for example Ash *et al.*, 1967) is used to compute the propagation times from the telescope to the barycenter. Compensation for two other effects is necessary to obtain arrival times in an inertial reference frame. First, delays due to interstellar dispersion (see p. 128) must be removed; in computing this term one must use the frequency at which the signal propagates through the interstellar medium rather than the observed frequency (which has been Doppler shifted by the earth's motion). Second, the terrestrial clock times must be corrected for an annual variation in rate resulting from changing time dilation as the earth moves around its elliptical orbit.

Pulse arrival times at the solar system barycenter (or more accurately, in an inertial reference frame) are given by

$$t_b = t_s + (\mathbf{r_s} \cdot \mathbf{n})/c - D/v^2 + \Delta t_r, \tag{6-1}$$

where t_s is the observed pulse arrival time at the observing site, $\mathbf{r_s}$ is the vector from the solar-system barycenter to the site, $\mathbf{n}$ is a unit vector in the direction of the pulsar, D is the dispersion constant (see Eqn. 7-4, p. 129), v is the observing frequency corrected to the barycentric frame, and Δt_r is the relativistic clock correction. This latter term may be expressed in terms of the mean anomaly, l, and eccentricity, e, of the earth's orbit as follows (Clemence and Szebehely, 1967; Blandford and Teukolsky, 1976):

$$\Delta t_r = 0.001661 \left[\left(1 - \frac{1}{8}e^2\right) \sin l + \frac{1}{2} e \sin 2l + \frac{3}{8} e^2 \sin 3l \right]. \tag{6-2}$$

Systematic errors may be introduced into the computed barycentric arrival times by the correction terms in Equation 6-1. The term representing the propagation time to the barycenter has an annual sinusoidal variation with an amplitude of close to 500 s. Hence, an error of only 0.''1 arc in either the position of the pulsar or the orientation of the ephemeris coordinate system can introduce a sinusoidal offset in the computed arrival times of 250 μs amplitude. As described below, the pulsar position is normally solved for as part of the fitting process. Pulsar positions determined in this way often have an accuracy limited only by uncertainty in the orientation of the coordinate system, currently about 0.''1 arc. Errors in the computed barycentric arrival times will also be introduced if, for other reasons, the ephemeris does not correctly predict the position of the earth with respect to the solar system barycenter. Comparison of different ephemerides derived from planetary radar observations shows that such errors are less than about 10 μs over periods of five years or so.*

As described in Chapter 7 (p. 130), the dispersion measure for the Crab pulsar is known to vary by a few parts in 10^4; for several other pulsars upper limits on possible variations are about one part in 10^3. For a pulsar with a dispersion measure of 100 cm^{-3} pc, the total dispersion delay at a frequency of 1 GHz is about 400 ms. Therefore, at frequencies less than or

---

* A much larger systematic error arises because of the poorly known mass of Pluto, but this term does not change much over a few years.

around 1 GHz, significant error may be introduced into the computed barycentric arrival times if a constant dispersion measure is assumed. Changes in dispersion measure can be monitored and then compensated for if arrival times are obtained simultaneously at two or more widely spaced frequencies.

The accuracy of the relativistic clock-correction term, Δt_r, could be confirmed if independent position measurements with an accuracy of the order of $0\overset{''}{.}1$ arc were available for a number of pulsars. However, as mentioned below (and Chapter 5, p. 92), the binary pulsar provides a potentially more sensitive test of these predictions.

Once a series of barycentric arrival times have been obtained using Equation 6-1, improved values of the pulse period (P), or the pulsation frequency ($\Omega = 2\pi/P$), its derivative, and possibly other parameters, can be obtained by a least-squares fitting procedure. The pulsation frequency may be expanded as a Taylor series

$$\Omega(t) = \Omega_0 + \dot{\Omega}(t - t_0) + \frac{1}{2}\ddot{\Omega}(t - t_0)^2 + \cdots, \tag{6-3}$$

where Ω_0 is the frequency at time t_0. The pulse phase at time t is then given by

$$\phi(t) = \phi_0 + \Omega_0(t - t_0) + \frac{1}{2}\dot{\Omega}(t - t_0)^2 + \frac{1}{6}\ddot{\Omega}(t - t_0)^3 + \cdots, \tag{6-4}$$

where ϕ_0 is the phase at t_0. The subscript on barycentric arrival times (t_b) has been dropped for clarity.

The first step in the least-squares analysis of a set of barycentric arrival times is to compute the predicted phase at each time t using Equation 6-4, where Ω_0 and $\dot{\Omega}$ are initial estimates of the pulsation frequency and its derivative. (At this stage $\ddot{\Omega}$ is normally assumed to be zero; in many cases of short data spans $\dot{\Omega}$ can also be ignored.) Normally, t_0 is chosen to be the first arrival time, so $\phi_0 = 0$ also. Provided Ω_0 and $\dot{\Omega}$ are known with sufficient accuracy, the phase at any observed arrival time, $\phi(t)$, will be close to $2\pi n$, where n is an integer. The difference between ϕ and the nearest whole cycle, $(\phi - 2\pi n)/\Omega$, is known as the *residual*, $\mathcal{R}$, expressed in units of seconds. An error in the assumed pulsation frequency will result in residuals that increase linearly with time, while an error in the frequency derivative will result in a parabolic residuals curve. This effect is illustrated in Figure 6-1.

Provided a data span of at least a year is available, extremely accurate pulsar positions can be computed from the arrival-time data. From Equation 6-1, an error in the assumed source position results in a sinu-

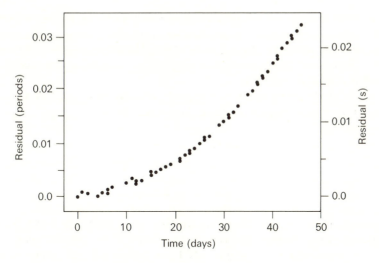

6-1 Pulse arrival-time residuals for PSR 0329 + 54, computed assuming that the period derivative ($\dot{P}$) is zero. Residuals are positive when the observed arrival time is later than the computed value. The data were recorded at the Five-College Radio Astronomy Observatory, Amherst, Massachusetts.

soidal term of period one year in the residuals. (Errors larger than a few arc minutes may result in incorrect pulse numbering, so a fairly accurate initial position is required.) If second-order and higher terms are omitted, the terms representing offsets in right ascension and declination can be separated and included explicitly in the least-squares fit of the residuals. If several years of data are available, proper motion of the pulsar may also be measured by including a time-dependent term in the position offsets.

The observed residuals can therefore be fitted by a function of the form

$$\mathscr{R} = \mathscr{R}_0 - \Omega_0^{-1}(t - t_0)\left[\Delta\Omega_0 + \frac{1}{2}\Delta\dot{\Omega}(t - t_0) + \frac{1}{6}\ddot{\Omega}(t - t_0)^2\right]$$

$$+ A[\Delta\alpha + \mu_\alpha(t - t_0)] + B[\Delta\delta + \mu_\delta(t - t_0)], \tag{6-5}$$

where $\Delta\Omega_0$ and $\Delta\dot{\Omega}$ are corrections to the assumed values of Ω_0 and $\dot{\Omega}$; $\ddot{\Omega}$ is the second time-derivative of the pulsar frequency; $\Delta\alpha$ and $\Delta\delta$ are corrections to the assumed values of right ascension and declination, respectively; and μ_α and μ_δ are the proper motions in right ascension and declination. The coefficients for the position-correction terms are

$$A = (r_E/c)\cos\delta_E\cos\delta\sin(\alpha - \alpha_E)$$

$$B = (r_E/c)[\cos\delta_E\sin\delta\cos(\alpha - \alpha_E) - \sin\delta_E\cos\delta], \tag{6-6}$$

where (α, δ) and (α_E, δ_E) are the 1950.0 coordinates of the pulsar and the earth, respectively, with respect to the solar system barycenter, and r_E is the distance from the earth to the barycenter. Higher-order terms in the Taylor expansion of the pulsation frequency could of course also be included if desired.

Values for the pulsar period and its derivatives may be expressed in terms of the pulsar frequency and its derivatives as follows:

$$P = 2\pi/\Omega$$
$$\dot{P} = -2\pi\dot{\Omega}/\Omega^2 \qquad (6\text{-}7)$$
$$\ddot{P} = 4\pi\dot{\Omega}^2/\Omega^3 - 2\pi\ddot{\Omega}/\Omega^2.$$

Accurate periods, period first-derivatives, and positions have been determined from timing observations for about 50 pulsars. Because the position correction terms (Eqn. 6-6) have an annual periodicity, and because the higher-order period derivatives are small, it is possible to obtain reasonably accurate period first-derivatives, even when the position is not well known, from just two measurements of the period made one year apart. Using this technique Lyne, Ritchings, and Smith (1975) obtained first-derivatives for another 36 pulsars. Periods, positions and, where known, period first-derivatives are listed in the appendix for all known pulsars. As described in the next section, a significant period second-derivative has been obtained for only the Crab pulsar.

Proper motion of a pulsar was first detected by Manchester, Taylor, and Van (1974) using pulse timing observations. Figure 6-2 is a plot of observed residuals for PSR 1133+16 showing the annual sine curve of linearly increasing amplitude resulting from proper motion. The proper motion derived from these results is $0.''32\pm0.''10$ arc per year, corresponding to a transverse velocity of 275 km s^{-1} for a distance of 180 pc. Proper motions have been independently measured for this and other pulsars by Anderson, Lyne, and Peckham (1975) and Backer and Sramek (1976) using interferometric techniques. These results and their implications will be discussed in Chapter 8.

For binary pulsars, pulse arrival times also reflect the orbital motion of the pulsar with respect to the binary system barycenter. These effects considerably complicate the timing analysis. For example, for a normal pulsar, an eight-parameter fit (Eqn. 6-5) is usually adequate to describe the systematic variations in the residual curve. For a binary pulsar, however, as many as 18 parameters are required. A detailed analysis of the pulse arrival-time variations expected for pulsars in binary systems has been given by Blandford and Teukolsky (1976). Below we give an outline

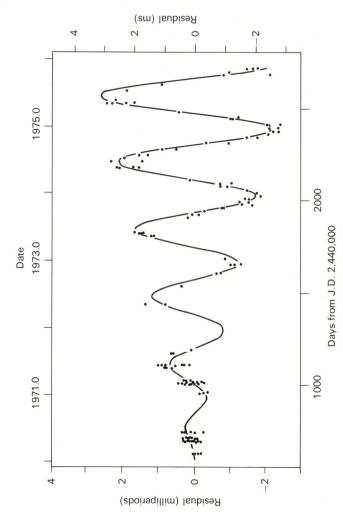

6-2 Arrival-time residuals for PSR 1133 + 16 showing the linearly increasing sinusoid resulting from proper motion of the pulsar.

of this analysis as described by Taylor and co-workers (1976). The results obtained for the binary pulsar PSR 1913 + 16 and some of their implications have been described in Chapter 5 (p. 91).

Since the semimajor axis of the PSR 1913 + 16 orbit is about 2.5 light seconds in size, the orbital motion introduces a variation in arrival times with an amplitude many times the pulsar period (0.059 s). Consequently, the orbital motion must be taken into account when computing residuals for the least-squares fitting procedure. The relation used for predicting the pulse phase is

$$\phi(t) = \phi_0 + \Omega_0(t - T_0) + \frac{1}{2}\dot{\Omega}(t - T_0)^2$$

$$+ \Omega_0 Q(2\pi RSP_b^{-1} - 1) - \dot{\Omega}Q(t - T_0), \tag{6-8}$$

where

$$Q = fg + h \sin E$$

$$R = -g \sin E + h \cos E$$

$$S = (1 - e \cos E)^{-1}$$

$$f = \cos E - e \tag{6-9}$$

$$g = x \sin \omega$$

$$h = (1 - e^2)^{1/2} x \cos \omega.$$

In these equations P_b, e, and $\omega = \omega_0 + \dot{\omega}(t - T_0)$ are initial estimates of, respectively, the binary orbital period, the orbital eccentricity, and the longitude of periastron (see Figure 5-8, p. 93); $x = a_1 \sin i$ is the projected semimajor axis; and E, the eccentric anomaly, is defined by

$$2\pi(t - T_0)/P_b = E - e \sin E, \tag{6-10}$$

where T_0 is a reference time of periastron passage. The pulsar frequency Ω_0 and the reference longitude of periastron ω_0 are also defined at this epoch.

As in the normal analysis, the pulse phase $\phi(t)$ would be an integral multiple of 2π for all arrival times t if the assumed parameters were precisely correct and there were no measurement errors. In practice, corrections to the parameters are obtained by least-squares fitting of the observed residuals. Possible secular variations in the parameters x, P_b, and e may also be obtained. The function to be fitted to the residuals is

(in units of seconds)

$$\mathcal{R} = \mathcal{R}_0 - \Omega_0^{-1}[(t - T_0) - Q]\left[\Delta\Omega_0 + \frac{1}{2}\Delta\dot{\Omega}(t - T_0) + \frac{1}{6}\ddot{\Omega}(t - T_0)^2\right]$$

$$+ A[\Delta\alpha + \mu_\alpha(t - T_0)] + B[\Delta\delta + \mu_\delta(t - T_0)]$$

$$+ c^{-1}[f \sin\omega + (1 - e^2)^{1/2}\cos\omega \sin E][\Delta x + \dot{x}(t - T_0)]$$

$$+ xc^{-1}[f \cos\omega - (1 - e^2)^{1/2}\sin\omega \sin E][\Delta\omega + \Delta\dot{\omega}(t - T_0)]$$

$$- 2\pi SRP_b^{-1}\left\{\Delta T_0 + (t - T_0)P_b^{-1}\left[\Delta P_b + \frac{1}{2}\dot{P}_b(t - T_0)\right]\right\}$$

$$- S[g(1 + \sin^2 E - e \cos E) - hf(1 - e^2)^{-1}\sin E][\Delta e + \dot{e}(t - T_0)]$$

$$+ \gamma \sin E. \tag{6-11}$$

The parameter γ combines the effects of variations in the transverse Doppler effect and gravitational redshift as the pulsar moves around its elliptical orbit, and is defined by

$$\gamma = \frac{2\pi a_1^2 e}{c^2 P_b}\left(2 + \frac{M_1}{M_2}\right). \tag{6-12}$$

The coefficients A and B for the position-correction terms are defined in Equation 6-6. Parameters obtained from an analysis of arrival-time data for PSR 1913+16 over 390 days were given in Table 5-2 (p. 93). Significant proper motion was not observed and would not be expected, as the distance indicated by the dispersion measure is about 5 kpc. As mentioned in Chapter 5 (p. 92), the time dependence of the coefficient of γ in Equation 6-11 is similar to that of Δx and $\Delta\omega$. After several years ω will have increased sufficiently to decouple these terms and hence allow a value for γ to be obtained.

Secular Period Variations

The frequency derivative ($\dot{\Omega}$) for all pulsars for which sufficient data are available has been found to be negative, i.e., the period is increasing with time. The first detection of this steady increase in period was for the Crab pulsar (Richards and Comella, 1969), for which $\dot{P} \approx 4.2 \times 10^{-13}$ s s^{-1} or about 36 ns per day. This pulsar, which has the shortest known period (33 ms), also has the largest known derivative. However, in general there is little correlation between period and period derivative (Figure 6-3). The smallest significant derivative measured so far is $\dot{P} \approx 1.9 \times 10^{-18}$ s s^{-1}, for PSR 1952+29, so the observed derivatives cover a range of over five

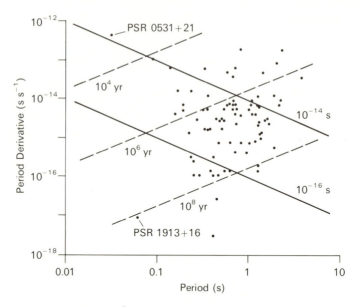

6-3 Period derivatives ($\dot{P}$) plotted against periods (P) for 83 pulsars. The solid lines are lines of constant $P\dot{P}$; pulsars will evolve to the right parallel to these lines if the braking index $n = 3$. The dashed lines are lines of constant characteristic age ($\tau = \frac{1}{2}P\dot{P}^{-1}$).

orders of magnitude. This range is much greater than the two orders of magnitude covered by the observed periods. The second-smallest derivative is for the binary pulsar PSR 1913 + 16, which has the second shortest period. Its derivative is over four orders of magnitude smaller than those of the Crab and Vela pulsars.

As will be described in more detail in Chapter 9, the observed regular increase in period is attributed to the loss of rotational energy and angular momentum via ejected particles and/or electromagnetic radiation at the rotation frequency. Except possibly for the recently observed γ-ray emission from PSR 1747 − 46 and PSR 1818 − 04 (Chapter 5, p. 80), the observed pulses constitute a very small fraction of the total energy loss. In most theoretical models the braking torque is proportional to some power, n, of the rotation frequency:

$$\dot{\Omega} = -K\Omega^n \tag{6-13}$$

or the equivalent

$$\dot{P} = (2\pi)^{n-1}KP^{2-n}. \tag{6-14}$$

The parameter n is known as the *braking index*, and K is a positive constant. For braking by magnetic-dipole radiation or particle acceleration in a dipole field, Equations 9-4 (p. 176) and 9-17 (p. 180) show that $n = 3$. Thus, according to Equation 6-14, the parameter $P\dot{P}$ is constant and hence pulsars should evolve along lines parallel to the lines of constant $P\dot{P}$ drawn in Figure 6-3. The parameter $P\dot{P}$ is related to the strength of the magnetic field and the moment of inertia associated with the neutron star. Both the Crab and Vela pulsars have values of $P\dot{P}$ close to 10^{-14} s, corresponding to surface magnetic-field strengths of about 3×10^{12} G for an assumed moment of inertia of 10^{45} g cm^2. However, for the one known binary pulsar and many longer-period pulsars values of $P\dot{P}$ are much smaller, indicating larger moments of inertia or, more probably, lower magnetic-field strengths.

The *characteristic time*, $T = P/\dot{P} = -\Omega/\dot{\Omega}$, is a basic time scale for the loss of rotational energy, and in most models is directly related to the age of the pulsar. This can be shown by integrating Equation 6-13. Providing $n \neq 1$, we obtain

$$t = -\frac{\Omega}{(n-1)\dot{\Omega}}\left[1 - \left(\frac{\Omega}{\Omega_i}\right)^{n-1}\right], \tag{6-15}$$

where Ω_i is the pulsation frequency at time $t = 0$, and Ω and $\dot{\Omega}$ are the current frequency and its derivative, respectively. If Ω is much less than Ω_i, the term in square brackets is approximately unity, and the time t becomes the *characteristic age*

$$\tau = -\frac{\Omega}{(n-1)\dot{\Omega}} = \frac{P}{(n-1)\dot{P}} \qquad (n \neq 1). \tag{6-16}$$

Hence, for constant n, the characteristic age represents an upper limit to the actual age of a pulsar. If the pulsar was born with Ω_i not much greater than the present Ω, the true age is less than τ. However, if the braking index has increased with time, then the present value of the characteristic age may underestimate the true age. A braking index of three is commonly adopted when computing characteristic ages, so $\tau = \frac{1}{2}P/\dot{P}$. Since the braking index is unlikely to be less than two, i.e., constant $\dot{P}$, this "dipole" characteristic age is, within a factor of two, a firm upper limit to the true age of a pulsar. Observational evidence suggesting that the true age is in general much less than the characteristic age is discussed in Chapter 8 (p. 161ff.).

Values of characteristic age for a braking index of three are shown in Figure 6-3; on this assumption, the characteristic age of the Crab pulsar is 1,240 years, somewhat larger than, but in reasonable agreement with, the actual age of about 920 years. From Equation 6-15, if the pulsation frequency of this pulsar has been decaying with $n = 3$ since birth ($t = 0$), then Ω_i was about 1.9Ω, i.e., a period of about 17 ms. In fact, the presently observed value of n is closer to 2.5 (see below), which suggests an initial period of about 22 ms. Of course, if the braking index has decreased during the lifetime of the pulsar, then the initial period may have been much less than these values. (The various factors affecting the observed braking index of pulsars are discussed in Chapter 9, p. 188ff.).

From the definition in Equation 6-13, the braking index is given by

$$n = \frac{\Omega\ddot{\Omega}}{\dot{\Omega}^2}, \qquad (6\text{-}17)$$

so, in principle, it can be computed directly from the pulsation frequency and its derivatives. To date, the only significant value obtained for a braking index is that obtained for the Crab pulsar. Groth (1975a) obtains $n = 2.515 \pm 0.005$ from an analysis of data recorded between March 1969 and April 1974. The quoted error (one standard deviation) reflects uncertainties resulting from the irregular period fluctuations—measurement errors are an insignificant contribution. Quite different values for the braking index are obtained from short data spans. For example, Boynton and co-workers (1972) obtained values ranging from 2.2 to 2.6 for several different data blocks each about four months long, and Nelson and co-workers (1970) obtained values ranging from 0.2 to 5.1 from several blocks each about 25 days long. Groth (1975a) shows that all of these variations can be attributed to the noise process responsible for the irregularities; the data are consistent with a smooth secular variation with constant braking index and superimposed random irregularities. These random fluctuations are discussed further in the following section.

Measurement of the period or frequency third-derivative would allow a check on the form of the braking law (Eqn. 6-13):

$$\dddot{\Omega} = \frac{n(2n - 1)\dot{\Omega}^3}{\Omega^2}. \qquad (6\text{-}18)$$

So far, no significant measurement of this parameter has been made. For the Crab pulsar, approximately 10 years of timing data will be required for a significant result (Groth, 1975a).

There is essentially no information on the braking index for other pulsars. For some pulsars period irregularities are large enough to mask any cubic term in the residuals, whereas for others the data spans are not yet long enough. With current methods of determining pulse arrival times, observations over 5–10 years or more will be required before a significant value of $\ddot{\Omega}$ can be determined.

If all pulsars followed the same or similar evolutionary tracks, then the braking index could be obtained from a plot of P versus $\dot{P}$. However the wide scatter of points in Figure 6-3 shows that all pulsars do not evolve along the same path. There are several possible explanations for this wide scatter. If pulsars do in fact evolve with a braking index close to three (as suggested by the Crab data and theoretical models), then Figure 6-3 implies a range of about five orders of magnitude in the quantity $B_0^2 R^6/I$, where B_0 is the surface magnetic field strength, R the neutron star radius, and I the moment of inertia (see Eqn. 9-5, p. 176). Thus, the observed range of $P\dot{P}$ implies that pulsars are born with (1) different masses, with R^6/I covering a wide range; (2) different magnetic fields; (3) similar magnetic field strengths that decay significantly over time scales of $\sim 10^6$ years; or (4) any combination of these. By assuming constant values of R^6/I and time scales for magnetic field decay of between 10^5 and 10^7 years, Lyne, Ritchings, and Smith (1975) have computed families of evolutionary paths on the P–$\dot{P}$ diagram (Figure 6-3). Initially, pulsars evolve parallel to the lines of constant $P\dot{P}$, but as the field decays the derivative decreases and the evolutionary paths become asymptotically vertical. Lyne, Ritchings, and Smith suggest that the observed number of short-period pulsars with small derivatives is less than would be expected if pulsars followed straight evolutionary paths with $n \approx 3$. The observed distribution is consistent with magnetic decay on a time scale of about 10^6 years. Nevertheless, this conclusion depends heavily on just two pulsars, Crab and Vela (in the upper-left part of the P–$\dot{P}$ diagram, Figure 6-3), and has been somewhat weakened by the subsequent discovery of the binary pulsar, which has a short period and small derivative. Theoretical models (p. 173) also suggest that magnetic decay on these time scales is unlikely. On the other hand, arguments presented in Chapter 8 (p. 161ff.) suggest that pulsars must turn off on this time scale, and magnetic decay is one way in which this could occur.

Random Period Variations

The first observation to show that pulsar periods were subject to unpredictable variations was the detection in March 1969 of a large, apparently discontinuous, decrease in the period of the Vela pulsar (Radhakrishnan and Manchester, 1969; Reichley and Downs, 1969). Since that time two similar events have been observed for this pulsar and two or three much smaller discontinuities have been observed in the period of the Crab pulsar. In addition to these discrete events, analysis of pulse arrival-time data has shown that, superimposed on the secular variations described in the previous section, there are small but quite significant irregular fluctuations. This so-called "restless" behavior, originally revealed by the different values of the braking index for the Crab pulsar obtained from different data sets (e.g., Boynton *et al.*, 1969; Nelson *et al.*, 1970), is present in most pulsars and appears to be random and unpredictable. The observational characteristics of these two types of period irregularities are described here; their interpretation is discussed in Chapter 9 (p. 191ff.).

By far the largest of the observed period discontinuities (sometimes known as "glitches") are those for the Vela pulsar. In each of the three events observed so far, the period decreased by about 200 ns, a large change compared to the regular rate of increase of about 11 ns per day. Figure 6-4 illustrates these three period jumps in relation to the secular increase. In each case the period decrease occurred between successive observations separated by about a week; theoretical models suggest that the change actually occurs within minutes.

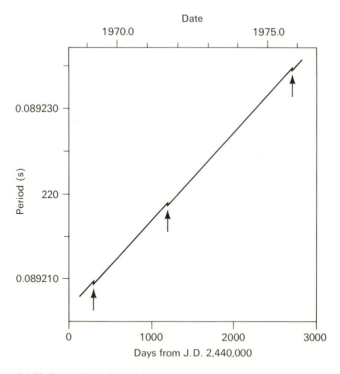

6-4 Variations in period of the Vela pulsar, PSR 0833 − 45, from 1968 to 1976, showing the three discontinuous steps in the regular increase of the period.

In each of these events the decrease in period was accompanied by an increase in period derivative. The fractional change $\Delta \dot{P}/\dot{P} = \Delta \dot{\Omega}/\dot{\Omega}$ was in each case about one percent, much larger than the fractional change $|\Delta P/P| = \Delta \Omega/\Omega$ of about 2×10^{-6}. For the first of these events (and probably also the other two), this increase in derivative decayed away after about a year, leaving the derivative at close to its original value.

The observed period discontinuities of the Crab pulsar are about a factor of 100 smaller than those of the Vela pulsar and are consequently much harder to separate from the random irregularities (described below). Two well-defined events have been observed so far, one in September 1969 (Boynton *et al.*, 1972) and another in February 1975 (Lohsen, 1975). There may have been a third event in October 1971 (Lohsen, 1972), but Groth (1975a) has argued that this event was not sufficiently large to be distinguished from a statistical fluctuation in the constantly present random irregularities. Unlike the Vela events, the Crab speed-ups varied considerably in size; for the 1975 event $\Delta \Omega/\Omega \approx 4 \times 10^{-8}$, about a factor of four larger than the 1969 event. As for the Vela pulsar, there was an increase in period derivative at the time of the period change, with $\Delta \dot{\Omega}/\dot{\Omega} \approx 2 \times 10^{-3}$. Figure 6-5 is a plot of the observed timing residuals for the Crab pulsar about the time of the 1975 speed-up, showing the abrupt change of slope resulting from the period decrease. These data show that, after the jump, the period relaxed back toward the extrapolated prejump data much more quickly than for the Vela pulsar. Figure 6-6 is a plot of the excess pulse frequency, i.e., the observed frequency minus the frequency obtained by extrapolation of the prejump data; the data show that the excess frequency decayed exponentially with a time constant of about 16 days. For the 1969 event the time constant was much shorter, about five days.

This exponential decay of excess frequency is consistent with the "two-component" model for neutron stars proposed by Baym and co-workers (1969). In this model, which is discussed in more detail in Chapter 9 (p. 192), the observed pulse frequency is determined by the rotation rate of the charged components of the neutron star (in particular the outer crustal layers). However, much of the mass, and hence moment of inertia, of the star consists of superfluid neutrons that are very weakly coupled to the charged components. Hence, a sudden change in the rotation rate of the charged components will be transferred to the neutrons very slowly and a relatively long time will elapse before equilibrium is restored. A fraction Q of the initial frequency jump ($Q \approx I_n/I$, where I_n and I are the moments of inertia of, respectively, the superfluid neutrons and the star as a whole) decays exponentially away with a time constant τ_d related to Q by

$$\tau_d = Q \, \Delta \Omega/|\Delta \dot{\Omega}|. \qquad (6\text{-}19)$$

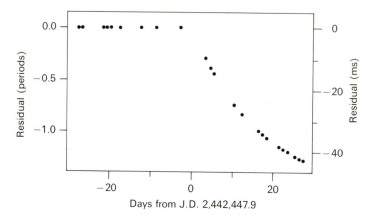

6-5 Timing residuals for the Crab pulsar, PSR 0531+21, about the time of the 1975 period discontinuity. A cubic polynomial was fitted to the arrival-time data before the jump. The nonlinear time dependence of the residuals after the jump indicates an increase in the period derivative. [After Lohsen, 1975.]

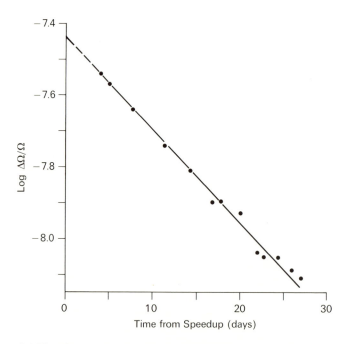

6-6 Plot of the fractional excess pulse frequency of the Crab pulsar immediately following the 1975 period discontinuity. The data are well fitted by an exponential law. [After Lohsen, 1975.]

Observed parameters for the Crab and Vela period discontinuities are summarized in Table 6-1.

In addition to sometimes exhibiting discontinuities, pulsar periods are continuously subject to small fluctuations—the so-called "restless" behavior. Detailed observations over long time scales have shown that, for most pulsars, arrival times are not precisely predictable. Any relatively short span of data (say, one to two months) can be fitted accurately by a cubic or, for the slower pulsars, a parabolic curve. For longer spans, however, significant residuals are often seen; furthermore the r.m.s. residual is larger for longer data spans. Figure 6-7 is a plot of residuals for the Crab pulsar over a five-year span after fitting a cubic polynomial to the data. Of course, by increasing the number of parameters fitted, the residuals can be reduced to low values. However, irrespective of the functional forms used, subsequent observations cannot be predicted accurately by such fits, and the number of parameters required increases with the length of the data span. For the five-year span of data shown in Figure 6-7, the r.m.s. residual is about one pulsar period or 33 ms. For comparison, the contribution to the pulse phase (Eqn. 6-4) from the linear term (Ω_0) is about 5×10^9 cycles, from the quadratic term ($\dot{\Omega}$) about 2×10^6 cycles, and from the cubic term ($\ddot{\Omega}$) about 2×10^3 cycles. Thus, although the irregularities are very significant in terms of the measurement errors, they are very small compared to the terms in the polynomial fit.

The unpredictable nature of the residuals led Boynton and co-workers (1972) to suggest that random processes were responsible for the observed variations. Three possibilities were considered: (1) a random walk in phase, e.g., changes in the emission-region location or the beam direction; (2) a random walk in frequency, e.g., changes in the stellar moment of inertia; and (3) a random walk in the first-derivative of the frequency, e.g., changes in the energy-loss processes. Several observational tests are available to determine which of these three random processes most closely describes the observed data. For the first process the r.m.s. residual would increase as $(t - t_0)^{1/2}$; for the second as $(t - t_0)^{3/2}$; and for the third as $(t - t_0)^{5/2}$, where $t - t_0$ is the total length of the data span. For each process the r.m.s. dispersion of the polynomial coefficients has a different dependence on $t - t_0$ and the spectrum of the fluctuations in the residual curve is different. Analysis of a two-year span of Crab pulsar arrival times by Boynton and co-workers (1972) showed that the data were most nearly consistent with frequency noise, i.e., a random series of small frequency jumps (of either sign) with the parameter $R\langle\Delta\Omega^2\rangle \approx 4 \times 10^{-21} \text{ s}^{-3}$, where R is the rate at which the frequency jumps occur and $\langle\Delta\Omega^2\rangle$ is their variance. Since the individual frequency jumps are not resolved with daily observations, $R \gtrsim 10^{-4} \text{ s}^{-1}$, and

TABLE 6-1

Parameters of observed period discontinuities in the Crab and Vela pulsars

	J.D. − 2,440,000	$\Delta\Omega/\Omega$	$\Delta\dot\Omega/\dot\Omega$	τ_d (days)	Q	Ref.*
Crab pulsar, PSR 0531+21	494.1 ± 0.9	$(9 \pm 4) \times 10^{-9}$	$(1.6 \pm 0.9) \times 10^{-3}$	4.8 ± 2.0	0.92 ± 0.07	1
	2447.4 ± 0.1	$(37.2 \pm 0.8) \times 10^{-9}$	$(2.1 \pm 0.2) \times 10^{-3}$	15.5 ± 1.2	0.96 ± 0.03	2
Vela pulsar, PSR 0833−45	280 ± 4	$(2.33 \pm 0.02) \times 10^{-6}$	$(8.1 \pm 0.2) \times 10^{-3}$	∼400	∼0.15	3, 4
	1192 ± 7	$(2.00 \pm 0.01) \times 10^{-6}$	$\sim 10^{-2}$	?	?	5
	2683 ± 3	$(1.97 \pm 0.01) \times 10^{-6}$	$(7.5 \pm 0.2) \times 10^{-3}$	455 ± 15	0.22 ± 0.01	6, 7

* REFERENCES: **1.** Boynton et al., 1972. **2.** Lohsen, 1975. **3.** Radhakrishnan and Manchester, 1969. **4.** Reichley and Downs, 1969. **5.** Reichley and Downs, 1971. **6.** Manchester, Goss and Hamilton, 1976. **7.** Manchester, Hamilton et al., 1976.

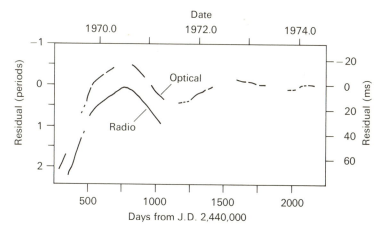

6-7 Timing residuals for the Crab pulsar, PSR 0531 + 21, over a five-year span. The residuals plotted are those remaining after fitting of a cubic polynomial to the data. Arrival times recorded prior to the end of 1971 were given lower weight in the fit and so have larger residuals. The upper curve is derived from optical measurements, the lower one (which has been arbitrarily displaced) from radio measurements. The two types of data generally agree to within better than 100 μs; the discrepancies can probably be attributed to uncertainties in the dispersion measure used to correct the radio data to infinite frequency. [After Groth, 1975a.]

hence the r.m.s. size of the frequency jumps must be $\langle\Delta\Omega^2\rangle^{1/2} \lesssim 6 \times 10^{-9}\ \mathrm{s}^{-1}$. Using a more sophisticated analysis based on orthonormal polynomials, which is valid even if the random process is nonstationary, Groth (1975a) found from five years of data that $R\langle\Delta\Omega^2\rangle = (2.1^{+0.9}_{-0.5}) \times 10^{-21}\ \mathrm{s}^{-3}$. There was no evidence for statistically significant time variations in either the polynomial coefficients or the strength of the random process.

Similar random irregularities are present in the periods of most other pulsars, with a few notable exceptions; no significant irregularities have been observed so far for the binary pulsar PSR 1913 + 16 (which has a period of 59 ms) or for several longer-period pulsars, including PSRs 1133 + 16, 1919 + 21, and 2016 + 28. Of four pulsars for which timing residuals were obtained over six years, two show significant fluctuations in period (Figure 6-8). For these two pulsars the residual curve has an overall cubic form because the third-order term is the first one omitted from the solution. However, this cubic term probably does not represent a secular change in the period derivative as in the Crab pulsar. Its size is dependent on the length of the data span fitted, and for the six-year span of data for PSR 1508 + 55 in Figure 6-8 the apparent braking index is

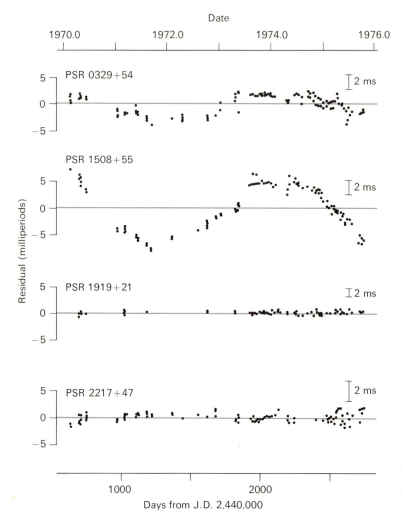

6-8 Timing residuals for four pulsars from data recorded by the authors and
D. J. Helfand at the National Radio Astronomy Observatory, Green Bank,
and the Five-College Radio Astronomy Observatory, Amherst. In each
case the residuals were plotted after fitting for the pulsar period, period
derivative, and position.

TABLE 6-2

*Root-mean-square timing residuals for seven longer-period
pulsars after fitting for period, period derivative, and position.
Pulsars are listed in increasing order of $\dot{P}P^{-5}$*

PSR	Data span (days)	P (s)	$\dot{P}P^{-5}$ $(10^{-15}\,\text{s}^{-5})$	$\langle\delta^2\rangle^{1/2}$ (ms)
1919+21	2020	1.337	0.3	≤0.3
1133+16	2085	1.188	1.6	≤0.3*
2016+28	2020	0.558	2.7	≤0.3
0329+54	2085	0.714	11.0	1.2
1508+55	2080	0.740	22.7	3.1
0823+26	1055	0.531	40.0	2.2
2217+47	2085	0.538	61.2	0.4

* After subtraction of proper motion terms.

unreasonably large (3,250). A preliminary analysis by Manchester and Taylor (1974) showed that, as in the case of the Crab pulsar, the irregularities in these longer-period pulsars have the characteristics of a random walk in pulsar frequency. An approximate value of $R\,\langle\Delta\Omega^2\rangle \approx 2 \times 10^{-27}\,\text{s}^{-3}$ was obtained from a four-year span of data for PSR 0329+54. Table 6-2 gives r.m.s. residuals for several longer-period pulsars after fitting for frequency, frequency derivative, and position. These residuals, along with those for the Crab pulsar, suggest that restless behavior is generally more pronounced in younger pulsars. The parameter $\dot{P}P^{-5}$, which is related to pulsar turnoff (see Chapter 10, p. 233), appears to be correlated with the size and/or frequency of the period irregularities.

7

Pulsars as Probes of the Interstellar Medium

The discovery of pulsars provided astronomers with a unique set of probes for the investigation of the interstellar medium. For example, the broad-band nature of pulsar emission makes possible the study of various absorption processes occurring in the galactic disk. At low frequencies free-free absorption by thermal electrons in the interstellar medium is potentially important, at least in some pulsars, and can yield information on the electron density distribution in the galaxy. Measurement of absorption at 1420 MHz by neutral hydrogen atoms gives information on HI cloud structure and, in many cases, provides an estimate of the pulsar distance. The pulsed nature of the signals makes them ideal for measurements of signal dispersion, and hence of the integrated electron content along the paths to the pulsars. When independent estimates of individual pulsar distances are available, the dispersions may be used to determine the mean interstellar electron density. Conversely, when a value for the mean density is assumed, individual pulsar distances may be derived from observed dispersion measures. Pulses frequently have a high degree of linear polarization which makes possible observations of Faraday rotation, from which the strength and direction of the interstellar magnetic field may be derived. The fact that pulsars are effectively point sources means that scattering by inhomogeneities in the interstellar electron distribution can alter the frequency-time structure of the signals observed at the earth. Measurement of the characteristic bandwidths and time scales of this structure may be used to infer some of the parameters of the scattering medium.

Interstellar Absorption

At frequencies below about 50 MHz the nonthermal galactic background radiation has a markedly reduced intensity near the galactic equator. This decrease is a result of free-free absorption by electrons in the interstellar medium. Dulk and Slee (1975) found in a sample of 15 supernova remnants lying within 45° of the galactic center at least eight with low-frequency spectral cutoffs at a frequency in the range 50–100 MHz, which the authors attribute to free-free absorption. The optical depth of this process is approximately proportional to v^{-2}, so the cutoff at low frequencies is quite sharp and easy to recognize. Pulsar emission is of course also affected by this type of absorption; as discussed in Chapter 2 (p. 24), low-frequency cutoffs are observed in the spectra of a number of pulsars. However, for most of these the cutoff frequency is probably too high to be caused by interstellar absorption alone. Part of the observed spectral turnover must be a result of low-frequency cutoffs intrinsic to the pulsar emission process. In addition, for the more distant pulsars, scattering processes (see p. 137ff.) broaden the pulses to such an extent that the pulsed flux is further reduced. These effects make interstellar free-free absorption difficult to detect in pulsars. For distant, long-period pulsars with intrinsically straight spectra, free-free absorption is likely to be significant at frequencies below about 100 MHz, particularly for sources in directions toward the galactic center.

Observations of the 1420 MHz spectral line of neutral hydrogen in absorption against continuum sources have led to a "two-component" model for the interstellar medium. The two components are (1) relatively cold, dense, and isolated regions known as clouds, and (2) a hotter, less dense intercloud medium. Since the optical depth is proportional to n_H/T_s, where n_H is the density of hydrogen atoms and T_s is the "spin" or excitation temperature for the 1420 MHz hyperfine transition, most of the observed absorption occurs in the cold, dense clouds. One of the principal problems in HI-absorption measurements against ordinary continuum sources is the determination of the emission contributed by neutral hydrogen within the antenna beam, the so-called "expected profile." Absorption measurements against pulsars are not troubled by this problem, because the expected profile can be determined during the portion of the period when the pulsar is "off." However, the fact that most pulsars have a low mean flux density means that most measurements of this type are sensitivity limited.

Out of a total of 43 pulsars for which hydrogen-line observations have been made, absorption has been detected in 28. Most of the pulsars having no absorption are relatively nearby and at galactic latitudes $|b| > 10°$. Emission and absorption spectra for two pulsars, PSR $0329 + 54$ and PSR $1933 + 16$, both of which are close to the galactic plane and reasonably distant, are shown in Figure 7-1. Where absorption is detected, an approximate distance can generally be inferred from the velocity extent of the absorption together with a model for the differential rotation of the galaxy. Since the systematic velocities are zero for all gas in directions toward the galactic center and anticenter, the method is of limited applicability in these directions. However, for the majority of pulsars it is the only method of determining distances independently of the dispersion measure, and so is very important. Distance estimates (or limits) determined in this way are listed for 26 pulsars in Table 7-1, together with distance estimates for two pulsars (Crab and Vela) based on observations of the associated supernova remnant. The upper limits are inferred from the absence of absorption at more negative or more positive velocities (depending on the quadrant of galactic longitude). Observations of absorption against continuum sources at low galactic latitudes (e.g., Radhakrishnan *et al.*, 1972) show that there are few breaks in absorption spectra at velocities where hydrogen is seen in emission. However, it is clear from the observed lack of absorption in about one-third of the pulsars studied that the absorbing medium is much more clumpy than the emitting medium, so these upper limits are not always very reliable. The derived distances are most reliable for those relatively high dispersion pulsars near the galactic plane and away from the directions $l = 0$ and $l = 180°$. For PSR $1933 + 16$ (Figure 7-1) absorption is seen out to the tangential velocity (~ 70 km s^{-1}), which indicates that the pulsar lies beyond the distance where galactic rotation is tangential to the line of sight. For some pulsars, e.g., PSR $1240 - 64$ and PSR $1859 + 03$, absorption is seen at both positive and negative velocities (Ables and Manchester, 1976). This shows that these pulsars are beyond the solar circle (the circle centered on the galactic center that passes through the sun) on the far side of the galaxy. The derived distance for PSR $1859 + 03$, 20 ± 2 kpc, is the largest known for any pulsar.

The fact that pulsars are effectively point sources of radio emission, together with the absence of the "expected profile" problem, makes them valuable probes of the distribution of interstellar hydrogen. Absorption

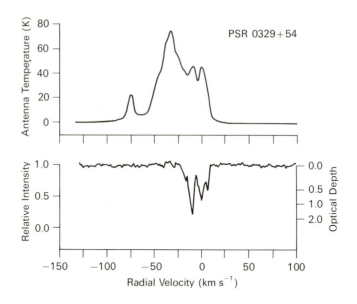

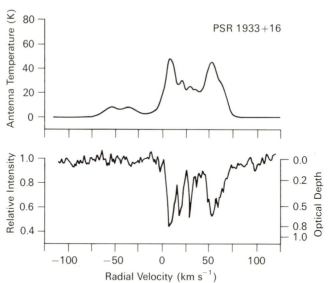

7-1 Profiles of emission (upper) and absorption (lower) by neutral hydrogen in the directions of the pulsars PSR 0329 + 54 and PSR 1933 + 16. The velocity resolution of the spectra is 1.7 km s^{-1} and radial velocities are with respect to the local standard of rest. [From Gordon and Gordon, 1973.]

TABLE 7-1

Mean electron densities along lines of sight to the 28 pulsars with independently determined distances

PSR	l	b	DM $(cm^{-3}\ pc)$	Distance (kpc)	Method	$\langle n_e \rangle$ (cm^{-3})	Ref.*
0138 + 59	129.1°	− 2.3°	34.8	∼ 3.0	HI	0.012	1
0329 + 54	145.0	− 1.2	26.8	2.3–2.9	HI	0.009–0.012	7
0355 + 54	148.2	+ 0.8	57.0	1.0–2.0	HI	0.028–0.057	1
0525 + 21	183.8	− 7.1	51.0	∼ 2.0	HI	0.026	2
0531 + 21	184.6	− 5.8	56.8	2.0	SNR	0.028	3
0736 − 40	254.2	− 9.2	161	1.5–2.5	HI	0.064–0.107	4
0740 − 28	243.8	− 2.4	73.8	1.5–2.5	HI	0.030–0.049	2
0833 − 45	263.5	− 2.8	69.1	0.5	SNR	0.138	5
0835 − 41	260.9	− 0.3	147.6	2.4–5.0	HI	0.030–0.062	6
1154 − 62	296.7	− 0.2	267	10.5–12.5	HI	0.021–0.025	4
1240 − 64	302.1	− 1.5	207	12–16	HI	0.019–0.025	4
1323 − 62	307.1	+ 0.3	320	5–11	HI	0.028–0.062	4
1557 − 50	330.7	+ 1.6	270	7–11	HI	0.025–0.039	4
1641 − 45	339.2	− 0.2	450	4.5–5.3	HI	0.085–0.10	4
1642 − 03	14.1	+ 26.1	35.7	0.15–0.17	HI	0.21–0.24	1
1718 − 32	354.5	+ 2.5	128	> 1.0	HI	< 0.120	2
1749 − 28	1.7	− 0.9	50.9	< 1.0	HI	> 0.051	2
1818 − 04	25.5	+ 4.7	84.4	< 1.5	HI	> 0.056	2
1822 − 09	21.5	+ 1.4	19.3	< 1.5	HI	> 0.013	2
1826 − 17	14.6	− 3.3	207	> 1.5	HI	< 0.138	1
1859 + 03	37.2	− 0.6	402	18–22	HI	0.020–0.022	4
1933 + 16	52.4	− 2.1	158.5	> 6.0	HI	< 0.026	2
1946 + 35	70.6	+ 5.0	129.1	> 8.5	HI	< 0.015	1
2016 + 28	68.1	− 4.0	14.2	> 1.0	HI	< 0.014	2
2020 + 28	68.9	− 4.7	24.6	> 2.0	HI	< 0.012	1
2021 + 51	87.9	+ 8.4	22.6	< 1.0	HI	> 0.023	2
2111 + 46	89.0	− 1.3	141.5	4–6	HI	0.024–0.035	1
2319 + 60	112.0	− 0.6	96	2.8–3.8	HI	0.025–0.034	7

* REFERENCES: **1.** Graham *et al.*, 1974. **2.** Gómez-Gónzalez and Guélin, 1974. **3.** Trimble, 1968. **4.** Ables and Manchester, 1976. **5.** Milne, 1970. **6.** Gordon and Gordon, 1975. **7.** Booth and Lyne, 1976.

features are always narrower than the corresponding emission features, which gives strong support to the "two-component" model of the interstellar medium. As yet, however, there are insufficient data available for a detailed analysis of the properties of the absorbing clouds.

In principle, the parameters of the cloud and intercloud regions can be investigated by comparing the ratio of the integrated optical depth for HI absorption and the dispersion measure for various pulsars. By assuming that these two regions were ionized by low-energy cosmic rays with the same ionization rate in both the cloud and intercloud regions, Hjellming, Gordon, and Gordon (1969) computed electron and gas densities for the two components and the fraction of the path occupied by each. However, more recent data (e.g., Boksenberg *et al.*, 1972; Shaver *et al.*, 1976) show that electron densities in dense clouds are very low, which would imply correspondingly low ionization rates in these regions. It seems therefore that pulse dispersion and hydrogen absorption occur in different regions of space, and hence that their ratio gives little information about the conditions in either. This is consistent with the wide range of values for the ratio of integrated optical depth to dispersion measure found by Gordon and Gordon (1973).

Interstellar Dispersion and the Galactic Electron Distribution

Because of the presence of free electrons in the interstellar medium, the group velocity of radio waves (v_g) is slightly less than the velocity of light and is a function of frequency. In standard texts on wave propagation in plasmas (e.g., Ginzburg, 1970) it is shown that, for a homogeneous isotropic medium,

$$v_g = c(1 - \omega_p^2/\omega^2)^{1/2}, \tag{7-1}$$

where ω_p is the plasma frequency and ω the wave frequency. The plasma frequency (in Gaussian units) is given by

$$\omega_p^2 = 4\pi n_e e^2/m, \tag{7-2}$$

where n_e is the electron density, and e and m are the charge and mass of the electron, respectively. For continuous signals the reduced velocity is of course not observable, but the pulsed nature of pulsar emission makes it possible to evaluate the degree of dispersion from the difference in pulse arrival times at two different frequencies. From Equation 7-1 one obtains, to the first order in ω_p^2/ω^2, a time delay

$$t_2 - t_1 = \frac{2\pi e^2}{mc}(\omega_2^{-2} - \omega_1^{-2})\int_0^d n_e\, dl, \qquad (7\text{-}3)$$

where d is the distance to the pulsar. The column density of electrons in the path to the source, $\int_0^d n_e\, dl$, is known as the dispersion measure (DM). The dispersion constant (D) is a directly measurable quantity and is defined by

$$D = (t_2 - t_1)/(v_2^{-2} - v_1^{-2}), \qquad (7\text{-}4)$$

where $v = \omega/2\pi$ is the radio frequency in Hz. It is related to the dispersion measure by $DM = 2\pi mcD/e^2$, or, when DM is expressed in the hybrid units cm^{-3} pc, by

$$DM\,(cm^{-3}\,pc) = 2.410 \times 10^{-16}\,D \quad (Hz). \qquad (7\text{-}5)$$

Observations over a wide range of frequencies show that, in general, the dispersion relation (Eqn. 7-4) is accurately obeyed. The insignificance of higher-order terms (Tanenbaum *et al.*, 1968) may be used to set weak upper limits on the electron density and magnetic field in the path ($n_e < 10^4\ cm^{-3}$, and $B < 10^{-3}$ G, respectively). For pulsars with "double" integrated profiles, the separation between the components is also a function of frequency (see Figure 2-4, p. 19), so the apparent dispersion cannot be precisely proportional to v^{-2} for the whole profile. Observations by Craft (1970) show that in such cases the dispersion delays are proportional to v^{-2} near the center of the profile. This suggests that the orientation of the radiation beam center is fixed with respect to the rotating neutron star and that the outer components are symmetrically located about this fixed direction.

Dispersion measures have been obtained for all but one of the known pulsars. Values range between 2.969 cm^{-3} pc (for PSR 0950+08) and 450 cm^{-3} pc (for PSR 1641−45) and are listed in the appendix. As is the case for pulsar periods, approximate dispersion measures are usually found as part of the discovery process. Using normal pulse-timing procedures (Chapter 6, p. 102) with observations at frequencies spaced by tens of MHz, dispersion measures can be determined with an uncertainty of ∼0.01 cm^{-3} pc, that is, to about a part in 10^4. More accurate measurements require a wider frequency range—particularly if it can be extended toward the low-frequency end—or more accurate timing measurements.

As mentioned in Chapter 3 (p. 48), Rickett, Hankins, and Cordes (1975) have detected a significant correlation between microstructure pulses recorded at 111 and 318 MHz from PSR 0950+08, making possible dispersion measurements as accurate as one part in 10^5 for this pulsar. The narrow components in the integrated profile of the Crab pulsar (Figure 4-6, p. 70) allow dispersion measurements of similar accuracy for this pulsar. However, in the case of the Crab pulsar there is the complication that at low frequencies the profile is severely broadened by interstellar scattering. Rankin and Counselman (1973) have devised a method of separating the effects of scattering and dispersion and have found significant variations in both the scattering parameters and the dispersion measure of the Crab pulsar over a two-year period. The dispersion measure was observed to increase in two steps, each of about one part in 10^4, the first in September–October 1969 and the second in June–July 1970. It is possible that the first step was associated with the September 1969 period discontinuity in the Crab pulsar (see p. 115), but at the time of the second step period irregularities were not significantly larger than normal. No significant variations in dispersion measure have been observed for any other pulsar.

The distribution of pulsar dispersion measures with respect to galactic latitude for all but one of the known pulsars is shown in Figure 7-2. Moderate to large dispersion measures ($DM \gtrsim 40$ cm^{-3} pc) are found

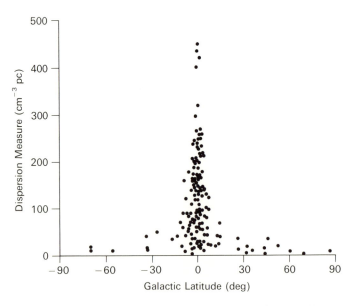

7-2 Plot of dispersion measure versus galactic latitude for 148 pulsars.

only at galactic latitudes $|b| \lesssim 10°$, which shows conclusively that the dispersion arises in the interstellar medium rather than in the immediate pulsar environment, and that pulsars belong to a galactic-disk population. The mean density of the dispersing electrons in the disk can be estimated because the distances to a number of pulsars are known from neutral hydrogen-absorption data (p. 125) or from a pulsar's association with a known supernova remnant. For 28 such pulsars the computed values of $\langle n_e \rangle = DM/d$ are listed in Table 7-1 and plotted against galactic longitude in Figure 7-3. The computed densities show considerable variation, but much of this is due to a few nearby pulsars for which the lines of sight pass through known HII regions. (For example, the large ionized region known as the Gum Nebula probably contributes at least half of the observed dispersion for PSR 0833 − 45.) Therefore, a reliable estimate of the average interstellar electron density is best based on the more distant pulsars, for which the fractional contribution to the dispersion measure of any intervening HII region is much less. For most of these pulsars the value of $\langle n_e \rangle$ is 0.03 cm^{-3} to within a factor of two, which therefore seems to be a good estimate of $\langle n_e \rangle$ over a large portion of the Galaxy.

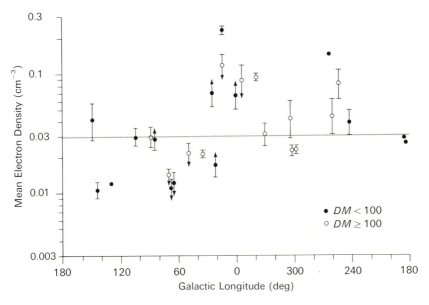

7-3 The mean interstellar electron density, determined from the dispersion measures of pulsars with independent distance determinations, and plotted against galactic longitude. The error bars represent the uncertainty in pulsar distance. For several pulsars no estimated errors were published; these points are therefore plotted without error bars but in general have uncertainties comparable to the other points. [After Ables and Manchester, 1976.]

Two systematic trends are suggested by the data in Figure 7-3. Firstly, the mean electron densities appear to be somewhat higher for most pulsars at longitudes within 30° of the galactic center, which suggests a larger mean density in the inner region of the galaxy. Secondly, for the longitude range 50°–80° low mean densities are observed. For distances up to about 8 kpc most of the path for this longitude range lies between the Orion and Sagittarius arms, which suggests that electron densities are lower in interarm regions.

The thickness of the distribution of free electrons in the z-direction (perpendicular to the galactic plane) has been deduced from measurements of free-free absorption (Bridle and Venugopal, 1969) and interstellar scattering of extragalactic sources (Readhead and Duffett-Smith, 1975). These authors find the equivalent half-thickness to be in the range 500–1000 pc. Information about the z-distribution of both the pulsars and the dispersing electrons can be obtained from a histogram showing the number of pulsars in equal intervals of $DM \sin |b|$, the "z-component" of dispersion measure (Figure 7-4). A simple and immediate conclusion is that the largest values of $DM \sin |b|$ are approximately 20 cm^{-3} pc, and therefore that $20/0.03 \approx 670$ pc represents a lower limit to the equivalent half-thickness of the electron layer.*

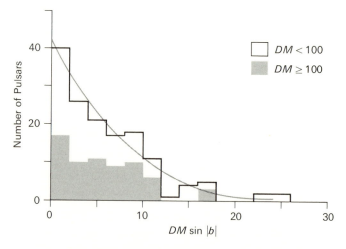

7-4 Distribution of the "z-component" of pulsar dispersion measures, $DM \sin |b|$. The superimposed curve is a least-squares fit to the data assuming exponential z-distributions of pulsars and electrons.

---

* One of the four pulsars with the largest $DM \sin |b|$, PSR 0736–40, is behind the Gum Nebula, which contributes substantially to its dispersion measure. However, the other three pulsars, PSRs 0450−18, 1541+09, and 2303+30, are probably free of such effects.

One can show that the expected form of the $DM \sin |b|$ histogram depends on the relative thickness of the pulsar and electron distributions in the following way. We assume that the density of thermal electrons is a function of z alone, $n_e(z)$, and that the number density of pulsars is similarly given by some function $N_p(z)$. If the z-component of dispersion measure is defined as

$$\Delta(z) \equiv DM \sin |b| = \int_0^z n_e(z') \, dz', \tag{7-6}$$

then the number of pulsars expected in equal intervals of Δ is given by

$$N(\Delta) = N_p[z(\Delta)] \, dz/d\Delta. \tag{7-7}$$

Furthermore, if explicit functional forms are assumed for $n_e(z)$ and $N_p(z)$, then $N(\Delta)$ can be evaluated and compared with the observed distribution in Figure 7-4. In principle it is possible to fit the assumed functions to the data using the method of least squares and thereby solve for functional parameters such as the scale heights. This was done for assumed exponential functions

$$n_e(z) = 0.03e^{-|z|/h_e} \tag{7-8}$$

$$N_p(z) = N_0 e^{-|z|/h_p}, \tag{7-9}$$

for which the expected shape of the histogram is given by

$$N(\Delta) = \begin{cases} N_0 \left[1 - \dfrac{\Delta}{0.03h_e} \right]^{(h_e/h_p)-1}, & \Delta < 0.03h_e \\ 0, & \Delta \geq 0.03h_e. \end{cases} \tag{7-10}$$

A least-squares fit of this relation to the data (Figure 7-4) yields $h_p = 230 \pm 20$ pc and $h_e \approx 1000$ pc (Taylor and Manchester, 1977). The latter value is in good agreement with results obtained by the other methods discussed above. The scale height of the pulsar distribution is discussed further in Chapter 8.

The thermal electrons causing pulsar dispersion, free-free absorption, and interstellar scattering are widely distributed throughout the interstellar medium, and it is of interest to attempt to understand the processes by which the required levels of ionization are maintained. High-density "classical" HII regions make a significant contribution to the dispersion measure for only a few pulsars. Possible ionizing agents for the distributed ionized medium include cosmic rays and soft X-rays (from supernovae or pulsars) and ultraviolet radiation (from O–B stars, central stars of planetary nebulae, pre-white dwarfs, or very old supernova remnants). Recent observations, especially of ultraviolet interstellar spectral lines (reviewed by Spitzer and Jenkins, 1975) are inconsistent with steady-state models involving ionization by cosmic rays or X-rays. Furthermore, it is unlikely that cosmic rays could propagate sufficiently far from their sources to generally ionize the interstellar medium (Wentzel, 1974). Time-dependent models involving X-ray bursts from supernovae (Gerola *et al.*, 1974) or emission from "UV stars" (Lyon, 1975) seem to be more consistent with the spectral-line observations. Neither ultraviolet radiation nor soft X-rays can penetrate the denser interstellar clouds, so most of the dispersing electrons are in the intercloud medium. UV stars are highly evolved and consequently widely distributed throughout the interstellar medium. In particular, they have a much greater scale-height than the O–B stars and so give a much larger scale-height to the electron layer than that characterizing either the O–B stars or the interstellar neutral hydrogen layer. Indeed, in the model of Lyon (1975) the electron density actually increases with distance from the galactic plane to reach a maximum at about 300 pc.

Faraday Rotation and
the Galactic Magnetic Field

In the weak magnetic fields of interstellar space the propagation of electromagnetic waves is "quasi-longitudinal" in essentially all directions, so normal modes of propagation are circularly polarized. Because of the slightly different indexes of refraction for these two modes (see for example Ginzburg, 1970), the plane of polarization of a linearly polarized wave rotates along the propagation path, an effect known as Faraday rotation. The angle of rotation after traversal of a path d is

$$\Delta\psi = \frac{2\pi e^3}{m^2 c^2 \omega^2} \int_0^d n_e B \cos\theta \, dl, \tag{7-11}$$

where B is the magnetic flux density and θ is the angle between the line of sight and the direction of the interstellar magnetic field. The rotation measure (RM) is then defined by

$$\Delta\psi = RM \, \lambda^2 \tag{7-12}$$

so that

$$RM = \frac{e^3}{2\pi m^2 c^4} \int_0^d n_e B \cos\theta \, dl. \tag{7-13}$$

The rotation measure is positive for fields directed toward the observer and negative for fields directed away.

From Equation 7-13 one can see that the rotation measure represents a mean value of the line-of-sight component of the magnetic field along the path to the pulsar, weighted by the electron density. For pulsars (and pulsars alone) the normalization factor is at least approximately known because dispersion measure is proportional to the integral of n_e and, as discussed in the previous section, n_e is believed to be reasonably constant over a large portion of the galaxy. Hence, the mean line-of-sight component of the magnetic field is given by

$$\langle B \cos \theta \rangle = \frac{\int_0^d n_e B \cos \theta \, dl}{\int_0^d n_e \, dl} = \frac{1.232 RM}{DM}, \tag{7-14}$$

where the numerical factor is computed for B in μG (microgauss), RM in rad m^{-2}, and DM in cm^{-3} pc. Because they provide direct measures of the field strength, pulsars are powerful tools in investigations of the interstellar magnetic field. A number of other factors also enhance the importance of pulsars in this respect. First, they are galactic objects whose approximate distance is known, so the structure of the field can be studied using pulsars at different distances. Second, the emission from pulsars is often highly polarized so rotation measures can be relatively easily obtained. Third, there is apparently no Faraday rotation intrinsic to pulsars. Evidence for this includes the complete absence of differential Faraday rotation across integrated profiles (i.e., the variation in position angle is the same at all frequencies), the absence of large rotation measures for pulsars at high galactic latitudes, and the similarity of rotation measures for pulsars that are close to each other in space.

At the time of writing, rotation measures have been determined for a total of 61 pulsars; these measurements are listed in the appendix. The corresponding mean line-of-sight magnetic field strengths (Eqn. 7-14) are plotted in galactic coordinates in Figure 7-5. There is clearly a strong predominance of negative rotation measures (fields directed away from us) in the hemisphere $l = 0°$ to $180°$, and of positive rotation measures in the hemisphere $l = 180°$ to $360°$. With the exception of the region around the galactic center, pulsars that do not follow this pattern are either at high latitudes or are more distant than about 2 kpc. This strongly suggests that the interstellar magnetic field within a few kiloparsecs of the sun is dominated by a uniform field directed toward about $l = 90°$, $b = 0°$. This conclusion is similar to that reached from studies of rotation measures of extragalactic sources (e.g., Gardner *et al.*, 1969). A least-squares fit of a uniform "longitudinal" field model to data for pulsars lying within 2 kpc of the sun, except those within 30° of $l = 0°$, gives a field strength of $1.7 \pm 0.3 \, \mu G$ directed toward $l = 90° \pm 14°$ (Manchester *et al.*, 1977).

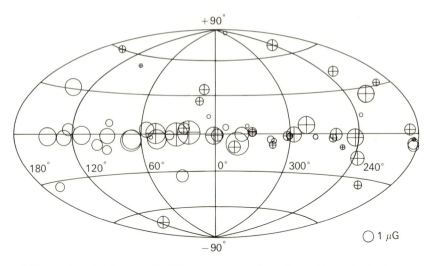

7-5 Mean line-of-sight magnetic field components in the paths to pulsars. For fields greater than 0.1 μG the area of the circle is proportional to the field strength (a circle representing a 1 μG field is shown in the lower right corner). A cross within a circle means the rotation measure is positive (field directed toward the observer). [From Manchester *et al.*, 1977.]

There are, however, some significant irregularities in this basically longitudinal field. For example, Figure 7-5 shows that field strengths in excess of 4 μG are observed for sources within 30° of the galactic center, where the projected component of a uniform field directed toward $l = 90°$ would be small. Also, there is considerable scatter in the field strengths observed in other directions: for the sources in the least-squares fit described above, the r.m.s. residual from the fitted longitudinal field is 1.3 μG. Observations of the nonthermal galactic background emission and the polarization of starlight also show clear evidence for irregularities in the local field. These observations tend to emphasize regions of increased field strength; for example, the intensity of synchrotron radiation is proportional to $B^{(\gamma+1)/2}$ (Eqn. 4-4, p. 64), where γ is typically about 3 for the galactic background. The North Polar Spur is a feature that is prominent in both galactic background and starlight polarization surveys. The large field component (4.4 μG) observed for the nearby pulsar PSR 1822−09 at $l = 22°$, $b = 1°$ almost certainly comes from strong fields associated with the Spur. These results suggest that, on the average, irregularities in the local field are comparable in strength to the longitudinal field.

For several pairs of pulsars close together in the sky (e.g., PSRs 1915+13 and 1929+10; PSRs 1727−47 and 1747−46) the rotation measure of the *nearer* source is consistent with the longitudinal field model, whereas that of the more *distant* source is of the opposite sign. Clearly there are large regions of the galactic disk where the field orientation is quite different to that of the local region. The much greater scatter observed in extragalactic rotation measures probably results, at least in part, from the path passing through several such regions.

Interstellar Scattering

Multifrequency observations of pulsars show that pulse-to-pulse fluctuations in intensity are correlated over wide frequency ranges, but that slower variations (with time scales of the order of tens of minutes) are correlated over much smaller bandwidths, generally less than 1 MHz. The type of complex frequency–time structure commonly observed is illustrated in Figure 7-6. Rickett (1969) first showed that the characteristic bandwidth of these spectral features is strongly dependent on dispersion measure, indicating that the structure is a result of propagation effects and not intrinsic to the pulsar. Subsequent work has shown that the type of spectral and temporal modulation shown in Figure 7-6 is satisfactorily accounted for by interference between rays traversing slightly different paths through the interstellar medium as a result of scattering by small fluctuations in the interstellar electron density. This process has been discussed by a number of authors; the following simple thin-screen model is based on the analysis of Scheuer (1968).

The phase perturbation of a ray passing through a fluctuation Δn_e in electron density of scale a is

$$\delta\phi = ae^2 \, \Delta n_e (mcv)^{-1}. \tag{7-15}$$

After propagation through a medium having a Gaussian distribution of density irregularities, with characteristic scale size a, the r.m.s. phase deviation of rays is given approximately by

$$\Delta\phi \approx (d/a)^{1/2} \, \delta\phi = (da)^{1/2} e^2 \, \Delta n_e (mcv)^{-1}, \tag{7-16}$$

where d is the distance between the source and the observer.

The scintillation effects resulting from propagation through this medium can be described approximately by considering that the scattering occurs in a thin screen situated midway between the pulsar and the observer, as illustrated in Figure 7-7. Scintillation of a point source with a modulation index near unity (in a narrow frequency band) will occur provided that (1) the scattering is strong, that is,

$$\Delta\phi > \pi, \tag{7-17}$$

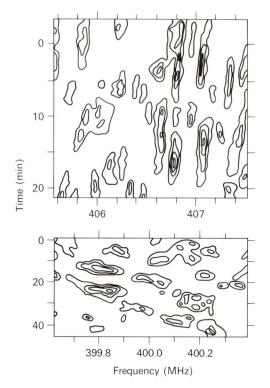

7-6 Pulse intensities of PSR 0329 + 54 (top) and
PSR 1642 − 03 (bottom) at frequencies near
400 MHz, plotted as contours in the
frequency–time plane.

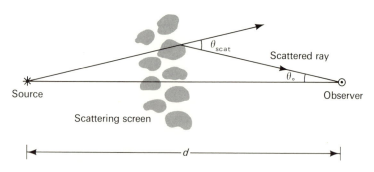

7-7 Schematic diagram showing the geometry of source, scattering region,
and observer.

and (2) the scattering is multiple so the observer sees several rays from the scattering region. The condition for this is

$$d\theta_0 \gg a, \tag{7-18}$$

where θ_0 is the angular extent of the scattered radiation as seen by the observer. For the geometry shown in Figure 7-7, the r.m.s. scattering angle is given by

$$\theta_{\text{scat}} \approx c\,\Delta\phi\,(2\pi a v)^{-1}, \tag{7-19}$$

and the apparent angular semidiameter of the source is

$$\theta_0 \approx \frac{1}{2}\,\theta_{\text{scat}} \approx \frac{1}{4\pi}\left(\frac{d}{a}\right)^{1/2}\frac{e^2\,\Delta n_e}{mv^2}. \tag{7-20}$$

In addition to angular broadening, the scattering process causes time smearing of an impulsive signal and both frequency- and time-dependent intensity scintillations. This is because a scattered ray received at a small angle θ suffers a time delay relative to the direct ray of $\Delta t \approx d\theta^2/2c$. For a Gaussian distribution of irregularities and hence a Gaussian distribution of scattering angles, the pulse shape is effectively convolved with a truncated exponential

$$g(t) = \begin{cases} \exp(-t/\tau_p) & t \geq 0 \\ 0 & t < 0, \end{cases} \tag{7-21}$$

where

$$\tau_p = d\theta_0^2/c \approx \frac{1}{ac}\left(\frac{e^2\,\Delta n_e}{4\,\pi m}\right)^2\frac{d^2}{v^4} \tag{7-22}$$

(see for example Cronyn, 1970). The scintillations are caused by interference between the direct and scattered rays, which remains constructive only over a limited bandwidth. Sutton (1971) has shown that if Δv is defined to be the frequency separation at which the correlation coefficient of observed intensity fluctuations drops to 0.5, then

$$\Delta v = \frac{1}{2\pi\tau_p} \approx 8\pi ac\left(\frac{m}{e^2\,\Delta n_e}\right)^2\frac{v^4}{d^2}. \tag{7-23}$$

As may be seen from Figure 7-6, time fluctuations in the observed signal strength are seen when the receiver bandwidth is less than Δv. These fluctuations result from the passage of the telescope through the diffraction pattern formed by the irregularity screen. For strong scattering the spatial scale of the pattern at the earth is

$$r_p \approx c(v\theta_0)^{-1}, \tag{7-24}$$

so the decorrelation time for the fluctuations is approximately

$$\tau_s \approx r_p/v \approx c(vv\theta_0)^{-1}, \tag{7-25}$$

where v is the velocity of the earth–pulsar line across the scattering screen. Since from Equation 7-20 $\theta_0 \propto v^{-2}$, we expect $\tau_s \propto v$. For weak scattering $\theta_0 \propto v^{-1}$, so r_p and hence τ_s are independent of frequency.

A more general model of interstellar scattering (or scattering in any inhomogeneous medium) assumes a medium whose properties are defined by the spatial covariance of the refractive index, or its Fourier transform, the spatial power spectrum. In the simple model discussed above this power spectrum is Gaussian, with its peak at a spatial frequency $q \approx 1/a$. However, similar results can be derived for other laws as well (see for example Rumsey, 1975; Lee and Jokipii, 1975). A power-law spectrum of density irregularities,

$$\Phi(q) = q^{-\beta}, \tag{7-26}$$

yields scaling rules similar to Equations 7-20, 7-22, 7-23, and 7-25 for the frequency dependence of observable parameters θ_0, τ_p, Δv, and τ_s. If we define $\kappa = \beta/(\beta - 2)$, then the scaling rules in the strong scattering case are

$$\theta_0 \propto v^{-\kappa} \tag{7-27}$$

$$\tau_p \propto v^{-2\kappa} \tag{7-28}$$

$$\Delta v \propto v^{2\kappa} \tag{7-29}$$

$$\tau_s \propto v^{\kappa-1}. \tag{7-30}$$

For the interstellar medium (and most other astrophysical plasmas) it is reasonable to expect $3 \lesssim \beta \lesssim 4$; in particular, the Kolmogorov turbulence spectrum gives $\beta = 11/3$, and thus $\kappa = 2.2$.

Observations have confirmed the basic theory outlined above, although aspects such as the form of the spatial power spectrum $\Phi(q)$ still await adequate testing. Mutel and co-workers (1974) have attempted to verify, by means of long-baseline interferometry, the frequency dependence $\theta_0 \propto v^{-2}$ predicted by the Gaussian model and Equation 7-20. The observed exponent over the range 26 to 115 MHz is in fact close to -2.0, but, as shown by Equation 7-27, a similar dependence is expected for a power-law spectrum with β close to 4. Thus, the spectrum of fluctuations in the interstellar electron distribution has yet to be determined observationally.

Direct observations of τ_p, Δv, and τ_s are generally in accord with the basic theory. For example, Backer's measurements of the decorrelation bandwidth (Δv) for the Vela pulsar (Figure 7-8) fit the v^4 law quite accurately. However, Ewing and co-workers (1970) find that for PSR 1133 + 16 the data are fitted by a line of slope 2.6 ± 0.6. For low-dispersion pulsars the multiple scattering condition (Eqn. 7-18) may begin to fail at the higher frequencies.

Equation 7-23 also predicts that the decorrelation bandwidth should be proportional to the inverse-square of distance d to the pulsar, which in turn is proportional to the dispersion measure, DM (p. 131). As shown in Figure 7-9, the observed dependence is approximately of this form, though a law proportional to $(DM)^{-3}$ fits the data somewhat better. This discrepancy suggests that scattering is more severe for higher dispersion pulsars. However, the observational data are probably biased by a lack of measurements of highly dispersed, weakly scattered pulsars.

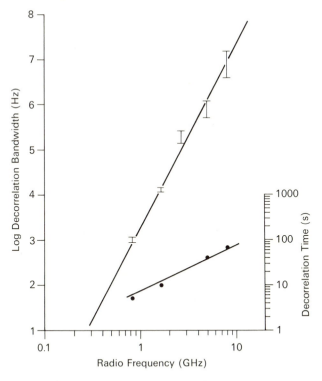

7-8 Decorrelation bandwidth (Δv) and decorrelation time (τ_s) for the Vela pulsar plotted as a function of frequency. The fitted line for the bandwidths has a slope of 4, and for the times a slope of 1. [From Backer, 1974.]

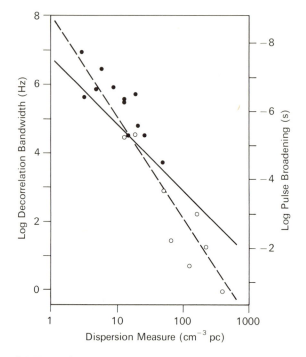

7-9 Decorrelation bandwidth (Δv) at 318 MHz plotted as a
function of dispersion measure. Filled circles represent
direct measurements of Δv; open circles are values derived
from measurements of pulse broadening (τ_p) using
Equation 7-23. Values measured at other frequencies were
scaled assuming a v^4 dependence. The solid line has a
slope of -2, the dependence expected from the theory;
however, the data fit the dashed line with a slope of -3
more closely. [After Sutton, 1971.]

Measurements of the dependence of τ_s on frequency are generally
consistent with the thin-screen model described above (see for example
Downs and Reichley, 1971). Data for the Vela pulsar are shown in Figure
7-8. From Equations 7-20 and 7-25 one should expect τ_s to vary approximately as $(DM)^{-1/2}$. In fact, Backer (1975) finds that the decorrelation
time is more nearly proportional to $(DM)^{-1}$, although there is considerable
scatter. Again, this suggests stronger scattering for the higher-dispersion
pulsars.

For the Gaussian model, the quantity $\Delta n_e a^{-1/2}$ may be computed
from Equation 7-23 if the characteristic bandwidth (Δv) and the pulsar
distance (d) are known. If, in addition, the frequency of transition between
weak and strong scattering (where $\Delta \phi \approx \pi$) is known, then $\Delta n_e a^{1/2}$ may

be obtained from Equation 7-16 and estimates of Δn_e and a obtained separately. Observations by Downs and Reichley (1971) at 2388 MHz suggest that for many of the stronger pulsars the transition frequency is about 1 GHz. Derived values of Δn_e are generally in the range $(1-5) \times 10^{-5}$ cm^{-3}, and of a in the range $(1-10) \times 10^{10}$ cm, although there are some notable exceptions. For example, Backer (1974) finds that for the Vela pulsar Δn_e is about a factor of 100 larger than these typical values. Similarly enhanced fluctuations are found in directions toward the galactic center. The mean electron density in the galactic disk is approximately 0.03 cm^{-3}, so the density fluctuations responsible for scintillation are quite weak. The value derived for a is not particularly meaningful as a physical scale size in the interstellar medium because the scintillation observations selectively emphasize a particular scale of irregularity equal to the size of the first Fresnel zone $(dc/2v)^{1/2}$.

Scintillations are observed only if the source diameter is less than the scale of the diffraction pattern at the earth, r_p. Otherwise, for a centrally located screen, patterns from different parts of the source overlap and cancel out. Light travel-time arguments (Chapter 10, p. 201) imply that, for pulsars, the source size is much less than typical values of r_p ($10^4 - 10^5$ km), so the condition is easily fulfilled. For other types of radio source, scintillation occurs only if the angular size of the source is consistent with the inequality

$$\theta_s \lesssim r_p/d \approx \frac{4\pi mcva^{1/2}}{e^2\,\Delta n_e d^{3/2}}. \tag{7-31}$$

Taking values of a and Δn_e as found from the pulsar observations and a path length through the scattering medium of 1 kpc, this relation shows that a source will scintillate at a frequency of 1 GHz only if its angular size θ_s is less than 10^{-5} or 10^{-6} arc s. Observations of compact extragalactic radio sources (Condon and Backer, 1975) show no evidence for scintillation of interstellar origin, implying that their diameters are greater than this limit.

Direct observations of pulse broadening have yielded results generally consistent with Equation 7-22, which predicts that the pulse shape is effectively convolved with a truncated exponential of time constant proportional to d^2/v^4. This relation shows that the broadening effect is much more pronounced at low radio frequencies and for the more distant, more highly dispersed pulsars. Profiles for PSR 1859+03 and PSR 1946+35 (DM = 402 and 129 cm^{-3} pc, respectively) at three frequencies are given in Figure 7-10 and show a pronounced tail of approximately exponential shape at the lower frequencies. Reliable measurements of both τ_p and Δv have been obtained for only a few

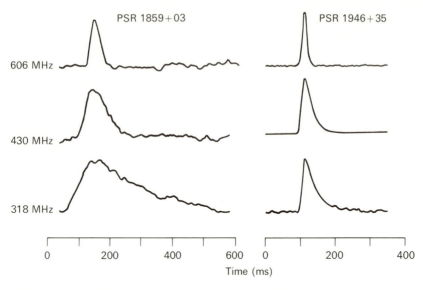

7-10 Integrated pulse profiles for PSR 1859+03 and PSR 1946+35 at three frequencies showing the exponential pulse broadening resulting from interstellar scintillation. [After Lang, 1971.]

pulsars, and never at the same observing frequency. For PSR 0833−45 Backer (1974) finds by extrapolation that, at a given frequency, $2\pi\tau_p \Delta\nu \approx 1$, in accordance with Equation 7-23.

Pulse broadening in the Crab pulsar, PSR 0531+21, has a dramatic effect on observed pulse shapes at frequencies below about 200 MHz because τ_p is the same order as the pulsar period, P. When τ_p is greater than P, the observed pulsed flux density is reduced, because the exponential tail of one pulse overlaps the rising portion of the next and effectively becomes part of the steady background. For $\tau_p \gg P$, Cronyn (1970) has shown that the pulsed flux is proportional to ν^4, so the scattering introduces a sharp low-frequency cutoff in the pulse spectrum. As described in Chapter 4, this cutoff occurs at about 100 MHz for the Crab pulsar.

Long-term monitoring of pulse shapes from the Crab pulsar has shown that both the amount of scattering and the shape of the scattering function change with time. From observations at several frequencies between 74 and 430 MHz Rankin and Counselman (1973) showed that a scattering function of the form $(t/\tau_1) \exp(-t/\tau_2)$, which represents scattering by two spatially separated thin screens, was consistent with the observations. Furthermore, the amount of scattering by one of these screens was variable on a time scale of weeks, suggesting that it is located within the Crab Nebula. For the steady screen, which is presumably associated with the interstellar medium, the parameter $v^4\tau_p$ had a value of about 6×10^5 s MHz^4. The second screen averaged about 12×10^5 s MHz^4 from May 1969 to August 1970, when within a period of four weeks it dropped to near zero. Further observations by Lyne and Thorne (1975) showed that during the first half of 1974 the amount of scattering increased dramatically, and reached a peak of about 800×10^5 s MHz^4 in November 1974. At several times the scattering function had two or more distinct peaks, these features persisting for at least several days. This implies that two or more discrete scattering regions separated by a substantial distance (accounting for the differential time delay) often existed. The large changes in scattering almost certainly occur within the Nebula, as implied electron densities and velocities are unreasonably large for a screen situated a large distance from the pulsar. For a screen in the vicinity of the wisps (see p. 68), electron density fluctuations of approximately 130 cm^{-3} are required to account for the observed scattering. Lyne and Thorne suggest that the large increases in scattering may be associated with the approximately coincident decreases in pulsed energy (see Figure 4-7, p. 73). There was no increase in dispersion measure associated with the increased scattering, so an increase in the relative amplitude of the fluctuations rather than an increase in the overall density of the scattering region must have been responsible. Variable scattering has not been observed for any other pulsar.

Close examination of the profiles of other pulsars where scattering is important shows that in most cases the leading edge of the profile is not as steep as would be expected if the scattering function were a simple exponential. In a comparison of observed pulse profiles with theoretical models, Williamson (1974) finds that they are fitted best by an extended screen filling not more than one-quarter of the path to the pulsar, or by two thin screens separated by less than one-eighth of the path length. Despite its intuitive appeal, a model in which the scattering medium fills the entire path between the pulsar and observer is excluded by the data. The amount of scattering does not seem to be correlated with the presence of known HII regions in the line of sight, but the strong scattering observed for the Crab and Vela pulsars suggests that the required fluctuations may occur preferentially in old supernova remnants.

Scintillation spectra such as those shown in Figure 7-6 are often quasi-periodic in form, especially for low-dispersion pulsars. This indicates that the number of independent rays reaching the observer is rather small. Wolszczan, Hesse, and Sieber (1974) used a computer simulation to estimate the number of such rays for several different pulsars. For PSR 0329 + 54 the number is about three, whereas for the distant pulsar PSR 1933 + 16, 10–20 rays are indicated. For many pulsars, the spectral features show a slow and systematic drift in frequency. For example, the patterns shown for PSR 0329 + 54 and PSR 1642 − 03 in Figure 7-6 drift at the approximate rates − 0.2 and + 0.4 kHz s^{-1}, respectively. In general, the drifting patterns last for only a limited time and may reverse in sense. They can be understood on the basis of the relative motion of the earth across an interference pattern formed by a few rays having different propagation delays (Ewing *et al.*, 1970). Computed pattern velocities are of the order of 100 km s^{-1}, which is not inconsistent with the pulsar velocities inferred from proper-motion studies (Chapter 8, p. 161).

If both the pulse-broadening time scale (τ_p)—or the decorrelation bandwidth (Δv)—and the decorrelation time (τ_s) are measured for a given pulsar, preferably at the same frequency, then the velocity (v) may be computed from Equations 7-22 and 7-25. Values obtained for different pulsars range from a few tens up to several hundred km s^{-1}. A more direct measurement of the scale size of the scintillation pattern (r_p) and its projected velocity across the earth can be obtained by cross-correlating the amplitude fluctuations observed at two or more well-separated sites. Galt and Lyne (1972) found that observed pattern delays for PSR 0329 + 54 were consistent with a velocity of 360 km s^{-1} directed approximately along the galactic plane. More recent interferometric measurements of proper motion imply a smaller space velocity of about 170 km s^{-1} for this pulsar. Rickett and Lang (1973) found that fluctuations of PSR 1133 + 16 were highly correlated at sites separated by ∼ 5000 km, showing that $r_p > 5000$ km, whereas the time delay was variable, even reversing in sign in one case. Pulse-timing observations (p. 106) and interferometric measurements (p. 161) show that this pulsar has a high space velocity. Obviously, neither the pulsar velocity nor the velocity of the interstellar medium could vary on such a short time scale, so it is clear that velocities determined from interstellar scintillation measurements using the simple approach outlined above are not very reliable. Slee and co-workers (1974) point out that if intrinsic intensity variations occur with the same time scale as the scintillation variations, the delays will be underestimated and hence the velocities overestimated. After compensating for this effect they obtain velocities of between 40 and 120 km s^{-1} for several pulsars. Velocities from scintillation observations that are too small or change sign indicate that the simple thin-screen theory is inadequate. Uscinski (1975) has shown that, for an extended medium containing transverse shear, the apparent velocities of pattern motion can be either too large or too small, and that relatively small changes in the path can produce large variations in the apparent velocity, including changes of sign. The discrepant velocities deduced from scintillation observations may possibly be accounted for in this way.

8

Pulsar Statistics and
Galactic Distribution

Without a doubt, many pulsars have yet to be discovered. They have eluded detection because their intrinsic luminosities are too low, they are too far away, their beamed emission does not sweep over the earth, or because of other observational selection effects resulting from the way in which pulsar surveys have been conducted. It would be very useful to know the frequency distributions that describe the relative number of pulsars as a function of period, height above the galactic plane, and distance from the galactic center, as well as the pulsar luminosity function. However, before we can discuss these distributions we must carefully evaluate the selection effects. It is, of course, also necessary to know the distances to pulsars if the distributions are to be meaningful. Therefore we shall begin this chapter with a discussion of pulsar distance determinations.

Pulsar Distances

In Chapter 7 (p. 131) we showed that the mean electron density, $\langle n_e \rangle$, is not too different from 0.03 cm^{-3} over a substantial portion of the galactic disk, and that the equivalent half-thickness of the layer of electrons is of the order of 1000 pc, considerably greater than the half-thickness of the pulsar distribution. With this information, we can estimate the distance of any pulsar from its dispersion measure (*DM*) and galactic latitude. If the electron density falls off exponentially in the *z*-direction with scale height h_e, then the distance to a pulsar is given by

$$d = \frac{-h_e}{\sin|b|} \ln \left[1 - \frac{DM \sin|b|}{h_e \langle n_e \rangle} \right] \tag{8-1}$$

Note that in the limit $DM \sin|b| \to 0$, the formula reduces to the simple relation

$$d = DM/\langle n_e \rangle, \tag{8-2}$$

which is an adequate approximation for many pulsars.

If an HII region in the line of sight contributes a substantial quantity (say *DM**) to a pulsar's dispersion measure, a better distance estimate will result if a "corrected" dispersion ($DM - DM*$) is used in Equation 8-1. Prentice and ter Haar (1969) have shown that values of *DM** can be calculated reasonably accurately for HII regions surrounding visible O–B stars or associations. The corrections typically amount to 5–15 cm^{-3} pc, and thus are an appreciable fraction of the total dispersion measure only for $d \lesssim 1$ kpc. Using Equation 8-1 and including the effect of HII regions within 1 kpc, we have computed the distances of all pulsars for which independent measurements are not available. The resulting values are listed in the appendix. Computation of dispersion-derived distances for those pulsars with independently measured distances (Table 7-1, p. 127) shows that three-quarters of the values agree to within a factor between 0.6 and 1.7. This indicates that the distance scale is statistically reliable, although a few individual distances may be in error by a factor of two or more.

Pulsar Searches and Selection Effects

About one-third of all known pulsars were discovered by methods involving direct pen recordings of receiver outputs on paper charts, followed by visual inspection of the charts for impulsive signals. Such methods do not make use of the periodic, dispersed nature of the signals being sought, and are broadly "tuned" to a single pulse width equal to the effective time constant of the recording equipment. Consequently, these methods do not approach the ideal sensitivities afforded by the antennas and receivers used.

A generalized survey of the sky for unknown pulsars is a complex five-dimensional process, because one does not know *a priori* the celestial coordinates (α, δ), period (P), dispersion measure (DM), or pulse width (W_e) of a new pulsar. When these quantities become known, further observations of the object become relatively easy, because post-detection signal-processing techniques can be used to optimize the system for the appropriate values of P, DM, and W_e. Therefore, an ideal pulsar-finding technique would include the capability of searching over as many combinations of α, δ, P, DM, and W_e as possible.

The desirable selectivity to a particular combination of period and pulse width can be achieved by Fourier analysis or by signal-averaging techniques (see for example Lovelace *et al.*, 1969; Staelin, 1969). Selectivity in dispersion can be obtained by using a multi-channel receiver and a means for delaying and recombining the detected outputs (Taylor, 1974). These requirements suggest the use of digital computers, which have been used to considerable advantage in the most recent pulsar surveys.

In all, successful pulsar searches have been conducted by 10 different research groups, using at least as many different methods, and observing at frequencies from 81.5 MHz to 1720 MHz. The selection effects that have determined the sample of known pulsars are thus far from uniform. However, 110 of the 149 known pulsars were first detected in one of three relatively extensive surveys made at the Molonglo, Jodrell Bank, and Arecibo observatories. Knowledge of the details of these three surveys, all of which were made at frequencies near 400 MHz, will therefore suffice to determine most of the relevant selection criteria.

Some of the particulars of the three surveys are summarized in Table 8-1. The Molonglo survey (Large and Vaughan, 1971) made use of the east–west arm of the Mills Cross antenna in a "split beam" mode, which produced two fan beams having half-power widths of $4°$ in declination and 1.5 in right ascension and separated by $3'$ in right ascension. Pulsar signals were recognized by visual inspection of chart recordings. Most of the sky south of $\delta = +20°$ was searched at least once, a given location being in each of the beams for $6/\cos\delta$ s. Most of the galactic plane at longitudes between $220°$ and $45°$ and latitudes less than $10°$ was searched again using a "dispersion remover" to extend the sensitivity to dispersions as high as 400 cm^{-3} pc. The minimum detectable flux density of this survey is a function of sky background temperature, period, and dispersion measure, but for typical values $P = 0.5$, $DM \leq 150 \text{ cm}^{-3}$ pc, and away from the galactic plane, the limit was approximately 80 mJy.

Several different methods of searching for pulsars have been used at Jodrell Bank, but the most successful technique has been that of Davies, Lyne, and Seiradakis (1972, 1973). The 76-m Mark IA telescope was used, together with an on-line digital computer, to search for periodicities in

TABLE 8-1
The three major surveys for pulsars, from which 110 of the 149 known pulsars have been discovered

	Molonglo	Jodrell Bank	University of Massachusetts
Antenna	Mills Cross	Mk IA	Arecibo
Sensitivity (K/Jy)	~5	~1.1	~14
Beamwidth	1.5×4.0	42'	10'
Search method	Chart records	Computer search for periods	Computer search for periods and dispersions
Sky coverage (steradians)	7	1.0	0.05
Coverage of galactic plane	$l = 220°$ to $45°$	$l = 352°$ to $115°$, $\|b\| < 9°$ $l = 115°$ to $250°$, $\|b\| < 4°$	$\left. \begin{matrix} l = 42° \text{ to } 60° \\ l = 182° \text{ to } 197° \text{ (part)} \end{matrix} \right\} \|b\| < 4°$
Receiver noise temperature, T_r (K)	600	110	130
Minimum flux density, S_0 (mJy)	80	15	1.5
Dispersion measure, DM_0, for which sensitivity is reduced by $\sqrt{2}$ (cm^{-3} pc)	150	250	1280
Pulsars discovered	31	39	40

the primary range 0.16 to 1.45 s. More than 5,000 beam areas (HPBW = 0°.7) were searched in the regions $-8° < l < 115°, |b| < 9°$ and $115° < l < 240°, |b| < 4°$, with about 11 minutes spent at each position. No use was made of the dispersed nature of the signals, but the 4-MHz bandwidth used was small enough that good sensitivity was maintained for dispersions up to about 250 cm^{-3} pc. The limiting flux density for this survey was about 15 to 30 mJy, depending on sky background temperature.

The most sensitive pulsar survey has been made by a group from the University of Massachusetts using the 305-m telescope at Arecibo (Hulse and Taylor, 1974, 1975b). Their method was to use a 64-channel receiver interfaced to a computer that carried out a three-dimensional search in period, dispersion measure, and pulse width. Nearly optimal sensitivity was achieved for periods $0.03 < P < 3.9$ s, dispersion measures $0 < DM < 1280$ cm^{-3} pc, and fractional pulse widths $0.016 < W_e/P < 0.125$. The nature of the Arecibo antenna restricted this survey to two small sectors of the galactic plane. Nearly complete coverage was made of the region $42° < l < 60°, |b| < 4°$, and about 24 square degrees were searched in the region $182° < l < 197°, |b| < 4°$. The limiting sensitivity was approximately 1.5 mJy.

Because the Molonglo pulsars were detected in what was basically a full-sky survey, they are relatively free of selection effects related to position in the sky. Despite the fact that this survey was more sensitive at high galactic latitudes where the sky background emission is weaker, more than half of the Molonglo pulsars have galactic latitudes less than 10°. This fact makes it clear that pulsars belong to the galactic disk population, and implies that confining the more sensitive pulsar surveys to low galactic latitudes has not resulted in very many pulsars being missed. This assertion is further confirmed by the distribution of pulsars with respect to height above the galactic plane (z) and projected distance from the sun (r), as shown in Figure 8-1. Most of the more distant pulsars in this sample were detected in the surveys at Jodrell Bank and Arecibo, which were confined to $|b| < 9°$ and $|b| < 4°$, respectively. The observed z-distribution of the distant pulsars is not significantly affected by these limited ranges of b; rather, it is essentially the same as the z-distribution of the nearby pulsars. Thus, the sample of known pulsars is only weakly affected by the limited latitude coverage of the high-sensitivity surveys.

The Molonglo and Jodrell Bank surveys had reduced sensitivity for pulsars with short period ($\lesssim 0.3$ s) and/or high dispersion ($DM \gtrsim 150$ cm^{-3} pc). On the other hand, the Arecibo survey maintained its full sensitivity up to $DM = 1280$ cm^{-3} pc, and was less sensitive by only about 3 dB for periods as short as 0.06 s. As shown in Figure 8-2, the Arecibo pulsars have a period distribution very similar to that for all

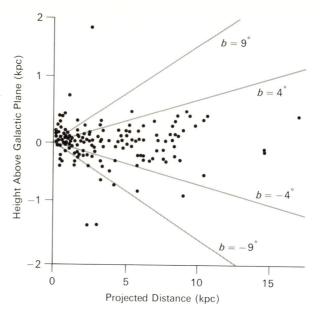

8-1 The distribution of known pulsars in height above
the galactic plane, $z = d \sin b$, and projected distance from
the sun, $r = d \cos b$, where d is the pulsar distance. The
Jodrell Bank and University of Massachusetts–Arecibo
pulsar surveys were confined primarily to the regions
$|b| < 9°$ and $|b| < 4°$, respectively, as indicated by the grey
lines.

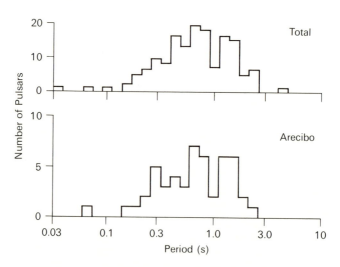

8-2 The distribution of periods of all 149 known pulsars (upper) and of
the 50 detected in the Arecibo survey (lower). Neither of the
histograms is strongly affected by observational selection effects;
within the statistical errors, they represent the true distribution of
pulsar periods.

known pulsars; in other words, the observed period distribution is not strongly affected by observational bias and must closely approximate the true distribution. The deficit of pulsars with periods near 1.0 s appears to be real, and suggests that there may be two distinct classes of pulsars. For example, most pulsars with periods less than one second may have been born in binary systems, and would thus have gained angular momentum during the mass-transfer stage (see p. 97).

Galactic Distribution of Pulsars

In this section we evaluate, first with semiquantitative arguments and then more rigorously, a set of functions that describe the distribution of pulsars with respect to luminosity and location in the galaxy. It is convenient to separate these distributions into three independent functions, $\rho_z(z)$, $\rho_R(R)$, and $\Phi(L)$, such that the total number of pulsars per unit area, projected on to the galactic disk at galactocentric radius R, is given by

$$D(R) = \rho_R(R) \int_{-\infty}^{\infty} \rho_z(z)\,dz \int_0^{\infty} \Phi(L)\,\frac{dL}{L}. \tag{8-3}$$

The use of three independent distributions is a valid approximation because correlations among the quantities R, z, and L are weak or nonexistent, as has been shown by Large (1971) and Seiradakis (1975). We choose to normalize the first two functions so that ρ_R (10 kpc) $= 1$ and $\int \rho_z(z)\,dz = 1$. Then $\Phi(L)$, which refers to the solar neighborhood, has the units pulsars per square kiloparsec per logarithmic luminosity interval.

First let us consider the pulsars in the immediate solar neighborhood. A projection on to the galactic plane of those pulsars within 1,500 pc of the sun is shown in Figure 8-3. One can see that, despite the fact that the sun is thought to be on the inner edge of the local spiral arm, nearby pulsars are somewhat more numerous toward the galactic center than in the opposite hemisphere and that the density of known pulsars decreases rapidly with increasing distance from the sun. We will argue below that the former effect is a real one and a consequence of a decreased density of pulsars at distances greater than 10 kpc from the galactic center. The apparent higher density of pulsars very close to the sun is of course not a real effect, but simply a reflection of the difficulty of detecting low-luminosity pulsars at large distances.

The nearby pulsars are distributed with respect to the galactic plane such that $\langle |z| \rangle = 153$ pc. This value probably underestimates the true local scale height of pulsars because a few of these objects were detected in low-latitude surveys that discriminated against nearby pulsars with large $|z|$. It is not unreasonable to infer that the equivalent half-thickness

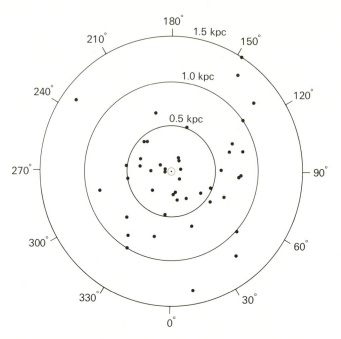

8-3 A projection onto the galactic plane of pulsars within 1,500 pc of the sun.

of the distribution of pulsars in the solar neighborhood is of the order of 200 pc.

If pulsars were uniformly distributed through a galactic disk of half-thickness 200 pc, we should expect $N(d)$, the number of pulsars with distance less than d, to increase as d^3 for $d < 200$ pc, and as d^2 for $d > 200$ pc. The observed $N(d)$ diagram for $d < 1500$ pc, shown in Figure 8-4, is in satisfactory agreement with this prediction up to $d \approx 500$ pc. At greater distances, an increasing number of pulsars have evidently been overlooked by the surveys. A total of 20 known pulsars lie within 500 pc of the sun, so a lower limit to the local density of active pulsars is 25 kpc^{-2}.

If the solar neighborhood is typical of other regions of the galactic disk, then the more sensitive pulsar surveys should detect more-distant pulsars with similar physical characteristics and distribution in z. In fact, such pulsars are observed. The distribution of all known pulsars with respect to z is shown in Figure 8-5. The mean distance of these objects from the galactic plane is $\langle |z| \rangle = 230$ pc. The fitted curve in Figure 8-5 represents an exponential distribution with scale height $h_p = 230$ pc, which was shown in Chapter 7 (p. 133) to be a good fit to the data. The fact that h_p

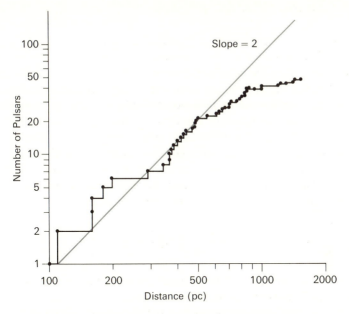

8-4 Number of pulsars with distances less than d, as a function of d, up to $d = 1,500$ pc. The sample becomes noticeably incomplete for $d \gtrsim 500$ pc.

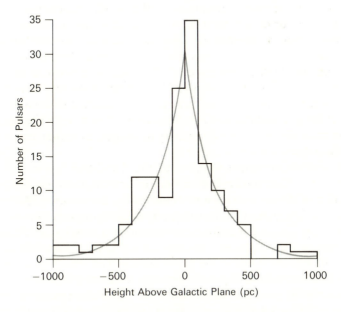

8-5 Distribution of all known pulsars with respect to height above the galactic plane, z. The fitted curve represents the best-fit exponential distribution, and has a scale height of 230 pc.

is not strongly dependent upon location in the galaxy is made clear by looking at values of $\langle |z| \rangle$ for samples of pulsars known to represent regions other than the solar neighborhood. For example, the 17 pulsars detected in the Arecibo survey with $DM \geq 200$ represent a region some 7 to 14 kpc from the sun, in direction $l \approx 50°$; for these pulsars, $\langle |z| \rangle = 250 \pm 40$ pc. Similarly, for the eight pulsars detected in the Jodrell Bank survey with $DM > 100$ and $|l| < 30°$, which represent a region approximately midway between the sun and the galactic center, the scale height is $\langle |z| \rangle = 260 \pm 60$ pc. In both cases there is a small observational bias against pulsars with low $|z|$ because the sky background radiation is stronger near $b = 0$. We conclude that the distribution

$$\rho_z(z) = \left(\frac{1}{460} \right) \exp \left[-\frac{|z|}{230 \text{ pc}} \right] \qquad (8\text{-}4)$$

is a good approximation to the normalized distribution of pulsars in the direction perpendicular to the galactic plane.

Several lines of evidence show that the density of pulsars is greater at galactocentric radii less than 10 kpc than outside this radius. One of these is the simple fact that far more than half of the known pulsars are found at longitudes within 90° of the galactic center. As mentioned above, this appears to be true even for the pulsars within 1 or 2 kpc of the sun, and one may infer that the density gradient with respect to galactocentric distance is already steeply negative at the solar distance. A similar conclusion may be reached by studying the longitude distribution of the pulsars detected in the Jodrell Bank survey (Lyne, 1974). The ratio between the pulsar density observed in directions toward the galactic center and that observed toward the anticenter is about three to one, with the pulsars in question lying typically at distances of 2 to 4 kpc from the sun. Thus the density of pulsars must increase with decreasing R, at least to within about 6 kpc from the galactic center.

Still another argument relevant to the distribution of pulsars with respect to R is illustrated in Figure 8-6, which plots pulsar flux densities as a function of distance (or, almost equivalently, dispersion measure). The diagram shows a smooth distribution at all dispersions up to about 300 cm^{-3} pc; only a few pulsars have larger dispersions, even though the University of Massachusetts–Arecibo survey, in particular, was far from sensitivity-limited at this value of dispersion. The most straightforward explanation of the lack of pulsars with flux densities greater than 1.5 mJy and dispersions greater than 300 cm^{-3} pc is that the Arecibo survey has penetrated to the edge of the galaxy, at least as far as the distribution of pulsars is concerned. The distance scale used to construct Figure 8-6

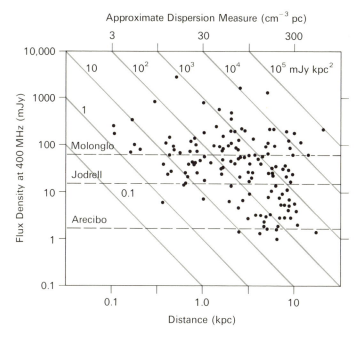

8-6 The distribution of all known pulsars with respect to flux density and distance from the sun. Approximate dispersion measures are indicated by the top scale. The dashed lines represent the approximate detection thresholds of the three principal pulsar surveys; the grey sloping lines separate decade-wide luminosity classes from 0.1 to 10^5 mJy kpc^2. The Arecibo survey maintained full sensitivity for dispersions up to 1,280 cm^{-3} pc, so the deficit of pulsars with $DM > 300$ cm^{-3} pc is a real effect.

suggests that this "edge" occurs about 10 kpc from the sun, or approximately 8.5 kpc from the galactic center. However, in the direction in question, $l \approx 50°$, the mean electron density in the galactic plane appears to be anomalously low (see Figure 7-3, p. 131); consequently the cutoff probably occurs closer to 15 kpc from the sun, or 11.5 kpc from the galactic center.

The relative density of pulsars as a function of galacto-centric distance, $\rho_R(R)$, and the pulsar luminosity function, $\Phi(L)$, can be evaluated quantitatively from the data obtained in the Jodrell Bank and Arecibo pulsar surveys (Taylor and Manchester, 1977). Each of these surveys was sufficiently systematic that it is possible to specify the minimum detectable flux density, S_{min}, as a function of galactic latitude and longitude and dispersion measure. To an adequate approximation, the detection limits are given by

$$S_{min} = S_0 \left(1 + \frac{DM}{DM_0}\right)^{1/2} \left(1 + \frac{T_{sky}}{T_r}\right), \quad (8-5)$$

where

$$T_{sky} = 30 + 270[1 + (l/40°)^2]^{-1}[1 + (b/3°)^2]^{-1} \, K \quad (8-6)$$

is the approximate 400 MHz sky background temperature at longitude l (where $|l| \leq 180°$) and latitude b; T_r is the receiver excess noise temperature; and S_0 and DM_0 are minimum-flux-density and maximum-dispersion parameters (tabulated in Table 8-1). For present purposes, a pulsar's 400 MHz "luminosity" is defined to be simply

$$L = S_{400}d^2, \quad (8-7)$$

where S_{400} is the mean flux density at 400 MHz in mJy and d is the distance in kpc.*

   With this information it is possible to compute the volume of the galaxy effectively surveyed for each interval of pulsar luminosity and galactocentric distance. By using the results already obtained for $\rho_z(z)$, volume can be converted to area projected onto the galactic disk. The luminosity function, $\Phi(L)$, and galactocentric distance dependence, $\rho_R(R)$, can then be evaluated by an iterative technique, as described by Large (1971).

   The results of such a calculation are illustrated in Figures 8-7 and 8-8. The straight line in Figure 8-7 corresponds to the equation

$$\Phi(L) = 350L^{-1.12} \, kpc^{-2}, \quad (8-8)$$

where the numerical coefficient is valid for semidecade luminosity intervals. This equation fits the data satisfactorily for luminosities ≥ 3 mJy kpc^2, although there appears to be a cutoff at lower luminosities. If $\Phi(L) = 0$ for $L < L_{min}$, the implied total density of detectable pulsars near $R = 10$ kpc is given by

$$D(10) = \int_{L_{min}}^{\infty} \Phi(L) \frac{dL}{L}, \quad (8-9)$$

and if we take $L_{min} = 3$ mJy kpc^2, we obtain $D(10) = 90 \pm 15$ kpc^{-2}. For $L_{min} = 10$ mJy kpc^2, the corresponding pulsar density is about 25 kpc^{-2}, a value consistent with the lower limit obtained above on on the basis of the local pulsar distribution alone.

---

   * For an assumed bandwidth of 400 MHz and a conical radiation beam of width 10°, the value $L = 1$ mJy kpc^2 is equivalent to 3.4×10^{25} erg s^{-1}.

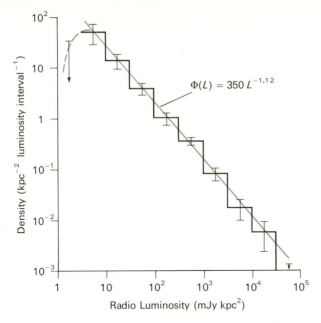

8-7 Pulsar luminosity function in the solar neighborhood. Densities are given in units of pulsars per square kiloparsec (projected onto the galactic plane) per semidecade of luminosity. The straight line corresponds to Equation 8-8. [From Taylor and Manchester, 1977.]

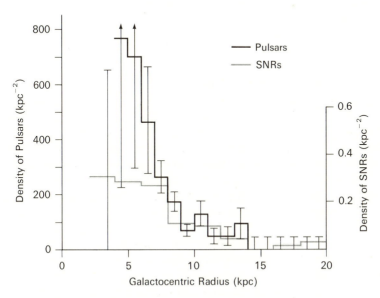

8-8 Density of pulsars (black line) and supernova remnants (grey line) as a function of distance from the galactic center. [From Taylor and Manchester, 1977; supernova data are from Clark and Caswell, 1976.]

As shown in Figure 8-8, the computed distribution $\rho_R(R)$ decreases sharply as R increases from 7 to 14 kpc, and is effectively zero for $R > 14$ kpc. Inside $R = 7$ kpc the distribution is rather poorly determined, though the data suggest that the density may peak near $R = 5$ kpc and decrease again toward the galactic center. It is possible that interstellar scattering (p. 137ff.) is much stronger in the innermost portion of the galaxy, which would imply a dependence of sensitivity on dispersion measure stronger than that assumed in Equation 8-5. Such an effect would help to explain the deficit of known pulsars at small R (Komesaroff *et al.*, 1973).

In principle, the total number of potentially observable pulsars in the galaxy can be determined from the integral

$$N_G = 2\pi \int_0^\infty RD(R)\, dR. \tag{8-10}$$

A numerical summation to approximate this integral, using the data in Figure 8-8, yields $N_G = (1.3 \pm 0.4) \times 10^5$ pulsars in the region $R < 14$ kpc. Because the area interior to $R = 7$ kpc is only one fourth of the total area, the fact that $D(R)$ is poorly determined in this region is not very important; any reasonable behavior affects N_G by less than 30 percent. However, other systematic errors could have larger effects. The biggest uncertainty in N_G arises from the poorly known value of the mean interstellar electron density. If the correct value were 0.02 cm^{-3}, rather than the assumed 0.03 cm^{-3}, then both $\Phi(L)$ and N_G would be smaller by a factor of about six. However, the estimate for N_G quoted above is in good agreement with the value of $(1 - 3) \times 10^5$ pulsars obtained in an independent analysis of the Jodrell Bank data by Davies, Lyne, and Seiradakis (1977) on the assumption that $n_e = 0.025$ cm^{-3}.

Pulsar Ages, Supernova Associations, and the Birth Rate

We now have available much of the information required for considering the class of pulsars as a galactic population. However, one important consideration, the ages and lifetimes of pulsars, remains to be discussed. As discussed in Chapter 6, timing observations can provide accurate measurements of a pulsar's period (P) and slowing-down rate ($\dot{P}$), from which a characteristic age can be computed ($\tau = \frac{1}{2}P/\dot{P}$). The characteristic age of the Crab pulsar, 1,240 years, is close to the 922 years elapsed since the supernova outburst; similarly, the characteristic age of the Vela pulsar, 1.1×10^4 years, is consistent with estimates of $(1 - 3) \times 10^4$ years for the age of the Vela supernova remnant obtained from semi-empirical relations involving linear diameter and surface brightness (e.g., Clark and Caswell, 1976). These two objects, which have the smallest characteristic ages of all known pulsars, both have periods <0.1 sec. These facts have been taken to indicate that pulsars in general are born with periods considerably smaller than their present periods, and that the characteristic ages of pulsars are good estimates of the true ages.

This assertion is not well supported, however, by pulsar ages and lifetimes estimated on a statistical basis using dynamical arguments. As discussed earlier (p. 133), pulsars have a z-distribution such that the scale height is about 230 pc; by contrast, the scale height of the O–B stars believed to be pulsar progenitors is ~80 pc (e.g., O'Connell, 1958) and that of supernova remnants only ~60 pc (Henning and Wendker, 1975; Clark and Caswell, 1976). Gunn and Ostriker (1970) have suggested that the wider pulsar distribution arises because pulsars receive a velocity impulse of order 100 km s^{-1} at birth. Consequently, in a typical lifetime of a few million years, a pulsar would move as much as a few hundred parsecs from the galactic plane, and the pulsar population would have a scale height of order 200 pc.

Direct measurements of proper motions have since confirmed that pulsars are indeed high-velocity objects. The available data are summarized in Table 8-2, arranged in order of characteristic age. Six of the nine pulsars listed have nearly the same characteristic age, between three and six million years. For these objects the mean transverse velocity is 214 km s^{-1}, the mean distance from the galactic plane is 153 pc, and the mean characteristic age is 4.3×10^6 yr. If the space velocities are isotropically distributed, we expect the mean z-component of velocity to be about

TABLE 8-2

Known proper motions and transverse velocities of pulsars

PSR	Characteristic age, τ (10^6 yr)	Proper motion, μ (10^{-3} arc s yr^{-1})	Distance, d (pc)	Transverse velocity (km s^{-1})	z (pc)	Ref.*
0531+21	0.0012	12 ± 3	2,000	110	−200	1
1237+25	2.3	102 ± 18	370	190	370	2
0834+06	3.0	52 ± 14	480	120	210	2
1929+10	3.1	159 ± 25	110	80	−7	3
1133+16	5.0	365 ± 36	180	310	160	2, 3, 4
0823+26	5.0	135 ± 10	790	510	420	2
0329+54	5.5	14 ± 4	2,500	170	−54	2
2016+28	5.9	20 ± 8	1000	95	−69	2
0950+08	17.3	≲100	100	≲50	71	3

* REFERENCES: **1.** Trimble, 1971. **2.** Anderson *et al.*, 1975. **3.** Backer and Sramek, 1976. **4.** Manchester *et al.*, 1974.

$214/\sqrt{2} \approx 150$ km s^{-1}. Therefore, if pulsars are born near $z = 0$, the observed $\langle |z| \rangle$ implies a mean age of just one million years. A more rigorous treatment, following Gunn and Ostriker (1970), assuming a velocity distribution that is Maxwellian in amplitude and random in direction and allowing for an initial scale height (h_i) of 80 pc, yields a mean "kinematic age" for these six pulsars of 1.0×10^6 yr. Uncertainty in the pulsar distance scale does not systematically influence this argument, because both the computed velocities and the z-distances are directly proportional to distance. The half-period for oscillation in the direction transverse to the galactic plane is about 10^8 years (Oort, 1965), much greater than the characteristic ages, so these pulsars cannot have returned to the plane after initially moving away. In any case, such oscillations are ruled out by the observed z-distribution of pulsars. Thus, the true age of these pulsars must be less than the characteristic age by a factor of about four.

A further test of this argument can be made by noting that if the velocities quoted in Table 8-2 are typical of pulsars as a class, then the value of $\langle |z| \rangle$ for any sample of pulsars is a monotonically increasing function of the mean age of the sample. The expected relation is graphed in Figure 8-9, again using the formalism presented by Gunn and Ostriker (1970). It is clear that a population of galactic objects moving in random directions, with mean speed $\sqrt{3}\langle |v_z| \rangle \approx 260$ km s^{-1}, cannot be older than a few million years and still have $\langle |z| \rangle$ as small as the 230 pc observed for pulsars. The data points in Figure 8-9 represent values of $\langle \tau \rangle$ and $\langle |z| \rangle$ for five groups of 13 to 19 pulsars each, which together comprise all pulsars with known characteristic ages. Unless the pulsars with $\tau \gtrsim 3 \times 10^6$ years have anomalously small velocities, they cannot be as old as their characteristic ages. Table 8-2 indicates only a weak tendency toward decreasing velocities with larger values of τ. We conclude, on the basis of this dynamical argument, that the true mean age of observed pulsars is one to two million years, and that the pulsars must "turn off" as they approach or exceed about twice this age. One would therefore expect that only samples with $\tau \lesssim 10^6$ years would exhibit a correlation between $\langle \tau \rangle$ and $\langle |z| \rangle$. This is in fact what is observed (Figure 8-9).

In Figure 8-10 we show a histogram of pulsar characteristic ages less than 2×10^7 years. Many pulsars have characteristic ages much greater than this value; for example, PSR 1952+29 has $\tau = 3.6 \times 10^9$ years. Since the characteristic age is a firm upper limit on the actual age of a pulsar (see p. 111), the rapid decrease in number of pulsars with increasing age shows immediately that many pulsars die after only a few million years. Following Davies, Lyne, and Seiradakis (1977) we can obtain an independent estimate of the lifetime of pulsars as follows. Of the 87 pulsars with known characteristic ages, about 20 have ages less than 10^6 years (Figure 8-10). Therefore, an upper limit on the "equivalent" lifetime

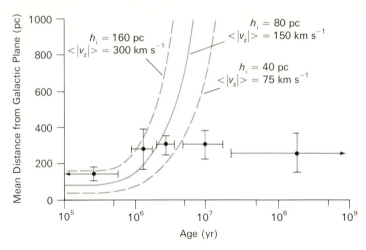

8-9 Expected mean distances from the galactic plane, $\langle|z|\rangle$, as a function of age for a population of objects derived from a parent population of scale height h_i. It is assumed that the second-generation objects receive at birth a velocity impulse that is random in direction and Maxwellian in amplitude, with mean speed in one coordinate equal to $\langle|v_z|\rangle$. The solid curve was computed on the assumptions that the scale height (h_i) of the parent population is 80 pc and that $\langle|v_z|\rangle = 150$ km s^{-1}; the dashed curves bracket these values by a factor of two in either direction. Points with error bars give the observed $\langle|z|\rangle$ for five groups of pulsars in different characteristic-age intervals, and show that very few pulsars can be older than a few million years. [From Taylor and Manchester, 1977.]

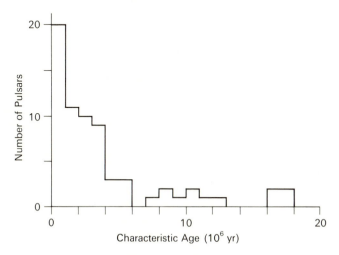

8-10 Histogram of the characteristic ages ($\tau = \frac{1}{2} P\dot{P}^{-1}$) of the 68 pulsars with τ less than 2×10^7 years. For the other 19 pulsars with known characteristic ages, τ is greater than 2×10^7 years.

(that is, the lifetime if all pulsars died at the same age) is approximately 87/20 or 4.5 million years. This result is in good agreement with the mean kinematic lifetime derived above.

It is clear that the larger characteristic ages (the tail of the distribution in Figure 8-10) are gross overestimates of the true ages. An actual age less than the characteristic age implies either that the pulsar was born with a period only slightly less than its current value, or that the braking index is (or has been) much greater than three. A number of authors, e.g., Gunn and Ostriker (1970), and Lyne, Ritchings, and Smith (1975), have suggested that substantial decay of the pulsar magnetic field occurs on a time scale of a few million years. This would have the effect of increasing both the braking index and the characteristic age and would also result in turnoff of the pulsed emission on the same timescale. If pulsars are born with relatively long periods (and a wide range of magnetic fields and/or masses, see Chapter 6, p. 113), then some other mechanism must be responsible for the turnoff in the pulsed emission (see Chapter 10, p. 234ff.). It should be noted, however that theoretical difficulties exist with both hypotheses. As will be discussed in Chapter 9 (p. 174), current models of neutron stars suggest that there is little decay of the magnetic field in less than about 10^7 years. Moreover, if O–B stars, which typically have radii of 3×10^6 km and rotation periods of about 10^5 s, are to produce neutron stars spinning as slowly as once a second, then a mechanism must be found for the star to shed most of its angular momentum either before, during, or shortly after collapse.

The arguments given above regarding the true ages of pulsars rely heavily on the fact that pulsars have typical space velocities in excess of 100 km s^{-1}. By what means are these high velocities produced? One possibility, discussed in Chapter 5 (p. 97), is that pulsars (or their progenitors) have in the past been members of close binary systems. In a binary system consisting of a 1-$M_\odot$ neutron star orbiting a 10-$M_\odot$ star with a period of one day, the neutron star has an orbital velocity of about 500 km s^{-1}. If the larger star were to explode as a supernova, ejecting more than half the original system mass beyond the orbit in less than one period, the system would become unbound, the neutron star flying off with a substantial fraction of its orbital velocity (Gott *et al.*, 1970). A second possibility is that pulsars are given high velocities at birth as a result of asymmetries in the supernova explosion. The kinetic energy of a 1-$M_\odot$ star moving at a velocity of 250 km s^{-1} is about 6×10^{47} ergs, a small fraction of the 10^{51} ergs thought to be released in a typical supernova explosion. Finally, pulsars may be accelerated to high velocities shortly after birth by the radiation-reaction force resulting from asymmetries in magnetic-dipole radiation (Tademaru and Harrison, 1975). This mechanism is discussed further in Chapter 9 (p. 177).

We now turn to a discussion of the origin of pulsars, and of the relationship between pulsars and supernovae. It is generally believed that neutron stars, and hence pulsars, are created in supernova explosions of massive stars. However, among the almost 150 known pulsars and approximately 120 known galactic supernova remnants (SNR), the evidence for associations on a one-for-one basis is not very strong. Only one pulsar, PSR 0531 + 21, is irrefutably associated with an observable supernova remnant, the Crab Nebula (see Chapter 4). The pulsar is located near the center of the remnant; the distance estimated from the dispersion measure, 1.9 kpc, is in good agreement with the estimated distance of about 2 kpc to the nebula; and, as mentioned above, the characteristic age of 1,240 years is only slightly greater than the time elapsed since the supernova outburst. Furthermore, the peculiar star identified by Baade and Minkowski in 1942 as the stellar remnant of the supernova has been shown to be the pulsar itself (see p. 68).

Evidence associating PSR 0833 − 45 with the Vela supernova remnant is not so overwhelming, but is probably sound. The pulsar lies within $0°5$ of the center of both the radio contours and the optically visible emission, and its rotation measure, 40 rad m^{-2}, agrees well with that of the radio continuum, 46 rad m^{-2}. The dispersion measure, 69 cm^{-3} pc, suggests a distance of about 2300 pc, but because more than half of the total dispersion is probably contributed by the Gum Nebula (Gordon and Gordon, 1975), the pulsar distance is likely to be less than 1000 pc. The distance of the SNR has been estimated as 375 to 1000 pc, based on a number of different arguments (see Kristian, 1970a for a summary). Finally, the pulsar's characteristic age is comparable to age estimates for the remnant.

The association of PSR 0611 + 22 with the supernova remnant IC 443 is much less certain. The pulsar lies just outside the shell-like radio contours, approximately $0°6$ from the center, and there is a suggestion of an indentation in the contours near the pulsar (Schönhardt, 1973). The pulsar dispersion measure is nearly twice that of the Crab pulsar, which suggests a distance of about 3500 pc. The distance to IC 443 has been estimated by Clark and Caswell (1976) as 2 to 2.8 kpc—although van den Bergh, Marscher, and Terzian (1973) have suggested that IC 443 is associated with the star HD 43836, which is only 500 pc distant. The characteristic age of the pulsar is 9×10^4 years, whereas the formulas of Clark and Caswell give an age for the radio remnant of only one-tenth this value. IC 443 has recently been detected as an X-ray source, and Winkler and Clark (1974) have argued that if it radiates by a mechanism similar to that of other known X-ray supernova remnants, its age cannot exceed 6,000 years. If this is correct, the angular separation between the pulsar and the center of the remnant would require an implausibly large velocity for the pulsar (3500 km s^{-1} for a distance of 2 kpc). Thus, the identification of PSR 0611+21 with IC 443 must be regarded as tentative at best.

A number of other possible pulsar–SNR associations have been proposed, but in no case is the evidence compelling. A summary of information regarding such associations is given in Table 8-3. At the top are listed the Crab and Vela supernova remnants; following these are six additional SNRs for which identification with a particular pulsar has been suggested. Although one or even two of these associations may be valid, we regard them as very doubtful statistically because the estimated distances and measured angular separations imply average velocities exceeding 1000 km s^{-1} for an assumed average SNR age of 3×10^4 years. The age figure is an upper limit and probably a conservative one; the supernova-remnant "completeness" statistics presented by Clark and Caswell (1976) strongly suggest that the average SNR age is not greater than about 2×10^4 years. We have argued above that the average pulsar age is between one and two million years. Therefore, in a steady-state situation not more than about two percent of the known pulsars are expected to be as young as the oldest supernova remnants. Consequently, it is not surprising that the vast majority of known pulsars are not surrounded by detectable supernova remnants.

We must also attempt to answer the question of why the vast majority of supernova remnants do not have detectable young pulsars embedded in them. The third group of SNRs in Table 8-3 are 13 for which a reasonably good observational limit on pulsed emission exists. The upper limits given for pulsed flux density are valid for dispersion measures $\lesssim 600$ cm^{-3} pc and periods $\gtrsim 40$ ms. The columns headed *DM* and Log *L* give the expected dispersion measure and the upper limit to pulsar luminosity, based on the distance quoted for the SNR. The angular separation between the SNR and its nearest known pulsar, listed in the last column of the table, are fully consistent with a "no association" hypothesis, given the density of known pulsars in the regions of sky in question. For most of the SNRs listed in Table 8-3 an undetected pulsar of low luminosity ($L \lesssim 100$ mJy kpc^2) is not excluded by the observations. According to the luminosity function illustrated in Figure 8-7, more than 90 percent of the pulsars in the galaxy have luminosities less than this limit, so the observational limits are not very significant. It is possible that with improved sensitivity many more pulsars will be found in supernova remnants. Of course, even if sensitivity were not a limiting factor, we would expect to observe pulsars in only about 20 percent of the known remnants because of beaming effects.

What can be said about the birth rate required to maintain the observed population of pulsars, given an estimated mean lifetime of only four million years? And how is the birth rate related to the rate of supernova

TABLE 8-3 *Evidence for associations between pulsars and supernova remnants*

Supernova remnants				Pulsars				
Galactic coordinates	Name	S_{400} (Jy)	d (kpc)	PSR	S_{400} (mJy)	DM (cm^{-3} pc)	Log L (mJy kpc^2)	Angular displacement
Certain or nearly certain associations								
G184.6−5.8	Crab Nebula	1300	2	0531+21	480	57	3.3	0.0°
G263.9−3.3	Vela X, Y, Z	2300	0.5	0833−45	2800	69	2.8	0.6
Questionable associations								
G189.1+2.9	IC 443	230	2	0611+22	130	97	3.2	0.6
G296.8−0.3		15	8	1154−62	110	267	4.1	0.2
G49.2−0.5	W51	200	4	1919+14	17	95	2.2	0.5
G55.6+0.7		15	6	1930+20	7	200	2.5	0.1
G55.7+3.4		1.6	9	1919+21	140	13	1.4	0.1
G89.0+4.7	HB 21	370	1.3	2021+51	130	25	1.9	3.9
No association*								
G5.3−1.1	A4	40	8		<60	240	<3.6	1.1
G6.5−0.1	W28	460	5		<60	150	<3.2	1.2
G21.8−0.5	Kes 69	110	6		<60	180	<3.3	1.9
G27.3+0.0	Kes 73	4	24		<60	720	<4.5	1.8
G34.6−0.5	W44	300	3		<60	90	<2.7	1.8
G43.3−0.2	W49B	50	11		<7	330	<2.9	0.5
G46.8−0.3		20	8		<4	240	<2.4	1.3
G53.7−2.2		12	5		<3	150	<1.9	0.9
G54.4−0.3		40	3		<6	90	<1.7	0.9
G111.7−2.1	Cas A	6700	3		<100	90	<3.0	1.6
G78.1+1.8	DR 4	100	6		<60	180	<3.3	7.6
G93.6−0.3	CTB 104A	50	2		<60	60	<2.4	4.7
G166.2+4.3	VRO 42.05.0i	10	4		<60	120	<3.0	20.9

* For this last group, the quoted value of DM is 0.03 times the supernova distance (in parsecs) and the last column lists the angular distance to the nearest known pulsar, even though it is not thought to be related.

occurrence? Earlier (p. 158) we showed that the density of observable pulsars in the solar neighborhood with luminosities greater than 3 mJy kpc^2 is about 90 kpc^{-2}. For a mean lifetime of 4×10^6 years this figure implies a pulsar birth rate of 2.2×10^{-5} yr^{-1} kpc^{-2} in the vicinity of the sun, or (after allowing for the radial distribution shown in Figure 8-8) one pulsar birth in the Galaxy every 30 years. If there is no large systematic error in the pulsar distance scale, this figure represents a lower limit for the pulsar birth rate required to maintain a steady-state population. If pulsars radiate in fairly narrow pencil beams, so that only about 20 percent are observable from the earth, the required local birth rate is about 10^{-4} yr^{-1} kpc^{-2}, or one every six years in the Galaxy. If the actual lower limit on pulsar luminosities is less than 3 mJy kpc^2 for pulsars of age less than 4×10^6 yr, then the birth rate may be even greater. On the other hand, if $\langle n_e \rangle = 0.02$ cm^{-3} the required birthrate drops to 3×10^{-5} yr^{-1} kpc^{-2}, or one pulsar every 40 years in the Galaxy.

Estimates of the rate of occurrence of galactic supernovae have varied widely. The largest rates quoted are approximately one supernova every 25 years, but estimates more typically suggest intervals of 50 to 150 years (see for example Tammann, 1974; Ilovaisky and Lequeux, 1972; Clark and Caswell, 1976). Thus, it seems that the rate of supernova explosions— at least of the type that leave long-lived remnants—may not be sufficient to explain the galactic population of pulsars. Another way of approaching this problem is to consider the death rate of moderate-to-high-mass stars, using star counts in the solar neighborhood and conventional stellar lifetime calculations. Ostriker, Richstone, and Thuan (1974) and Biermann and Tinsley (1974) have shown that the death rate of stars with mass greater than 2.5 $M_\odot$ is about 10^{-4} yr^{-1} kpc^{-2}, nearly the same as the computed pulsar birth rate if 80 percent of pulsars are unobservable because of beaming effects. On the other hand, the death rate of stars with mass ≥ 6 $M_\odot$ is smaller by a factor of five, and appears to be too small to explain the pulsar population. The scale height of very young pulsars (Figure 8-9) is more consistent with progenitors of mass about 2.5 $M_\odot$ than, say, that greater than 6 $M_\odot$. Unfortunately, the ultimate fates of stars in this mass range cannot yet be determined by numerical computation (Paczynski, 1973); it is not clear whether stars with masses as small as 2.5 to 4 $M_\odot$ can produce neutron stars. In any case, because the birth rate of observable supernova remnants is barely large enough to account for the observed numbers and lifetimes of pulsars, it seems important to consider the possibility that some, perhaps the majority, of pulsars are born in less spectacular circumstances than the explosive detonation of a high-mass star.

9

The Rotating
Neutron Star Model

The most widely accepted basic model for pulsars is the rotating neutron star model. The arguments that led to the general adoption of this model were summarized in Chapter 1. We shall begin this chapter by briefly reviewing our present understanding of neutron star structure. Since the discovery of pulsars and pulsating X-ray sources and their interpretation as rotating neutron stars, much work has been done on the structure and properties of these stars (extensive reviews of this work have been published by Cameron, 1970; Ruderman, 1972; Canuto, 1975a; and Baym and Pethick, 1975).

Pulsars are thought to have extremely strong magnetic fields, possibly the strongest anywhere in the universe. The presence of these strong fields on a rapidly rotating object results in the generation of strong electric fields and hence in the acceleration of charged particles to high energies. Analysis of the electrodynamics of the region surrounding the pulsar, the *magnetosphere*, is a complicated problem that has been (and continues to be) the subject of much investigation. Many models, involving various and often different approximations, have been developed. These theories and their principal conclusions will be reviewed in the second section of this chapter. Finally, in the third section, the processes that result in changes in the observed pulsar periods are discussed.

Neutron Stars

The concept of a neutron star—an object consisting primarily of neutrons, with mass about equal to that of the sun but radius of only about 10 km— was first discussed by Baade and Zwicky in 1934. Model calculations giving the structure of such an object were first made by Oppenheimer and Volkoff (1939). These authors considered a model in which the star is composed of noninteracting neutrons. At the enormous densities encountered in the interior of such stars, about 10^{14} g cm^{-3}, the neutrons form a degenerate Fermi gas; the pressure associated with this degeneracy is sufficiently large to prevent further collapse of the star. Subsequent analyses have considered more realistic compositions and equations of state for the various density regimes. A schematic illustration of the structure of a 1.33-$M_\odot$ neutron star computed from one of these models is shown in Figure 9-1.

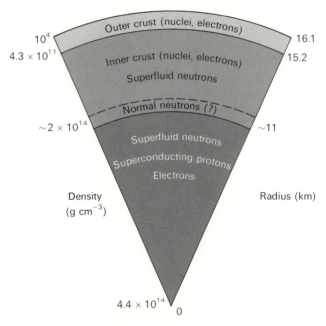

9-1 Cross-section of a neutron star having a total gravitational mass of 1.33 $M_\odot$. The equation of state on which this model was based is rather hard (i.e., relatively small density for a given pressure), so the central density does not reach 10^{15} g cm^{-3}, the density at which hyperon creation and neutron solidification may occur. [After Pandharipande *et al.*, 1976.]

In the outermost regions the structure of a neutron star is expected to be the same as in the interior of white dwarf stars: ^{56}Fe nuclei surrounded by a sea of degenerate electrons and a density at the surface of about 10^4 g cm^{-3}. The iron nuclei, because of their mutual electrostatic repulsion, are expected to form a body-centered crystalline lattice (provided the temperature is below about 10^{10} K), thereby giving the neutron star a solid outer crust. The Coulomb interaction energy is very high (~ 1 MeV per nucleus), so this crust will have a large shear modulus and hence be very rigid.

Beneath the outer crust of iron nuclei the density increases, resulting in an increased electron Fermi energy, which in turn results in electron capture by the nuclei. This layer therefore consists of neutron-rich nuclei, forming another crystalline lattice known as the inner crust and surrounded by a sea of degenerate electrons. At the interface with the outer crust, where the density is about 4.3×10^{11} g cm^{-3}, the electron Fermi energy is about 25 MeV and the nuclei begin to release free neutrons—the "neutron drip" point. As the density increases inward, the number of free neutrons increases rapidly until, at about 2×10^{14} g cm^{-3}, the nuclei completely dissolve into a neutron sea. The electron Fermi energy at this point is about 100 MeV. Within the inner crust the interaction between two neutrons is probably sufficiently attractive for the neutrons to form a superfluid (provided the temperature is below the critical value, about 10^{10} K). At densities higher than 2×10^{14} the material is thought to form a uniform sea of electrons, protons, and neutrons, with all but a few percent of the matter in the form of neutrons. The properties of matter at these super-nuclear densities are still the subject of much investigation, but it is expected that in this region the neutrons will again be superfluid and that the protons will be superconducting. There may be a zone of normal neutrons between the crustal and interior superfluid regions; the behavior of electrons remains normal throughout the star.

In some models, densities are in excess of 10^{15} g cm^{-3} in the core of the star. Electron and neutron Fermi energies become so great in such a core that new particles, such as muons and hyperons, are created. Although the characteristics of this region are very uncertain, calculations suggest that the first heavy particles to appear are the Σ^- and Λ° hyperons. Some authors, e.g., Canuto and Chitre (1973), have suggested that at densities greater than about 10^{15} g cm^{-3} the neutrons solidify into a crystalline lattice. However, in more recent work (e.g., Takemori and Guyer, 1975) no stable solid state can be found. Migdal (1973) has suggested that a solid state involving neutral pions may exist at these densities.

Neutron star models are computed by integrating the general-relativistic equation of hydrostatic balance (the Tolman-Oppenheimer-Volkoff equation),

$$-\frac{dP}{dr} = \frac{G[\rho(r) + P(r)/c^2][m(r) + 4\pi r^3 P(r)/c^2]}{r^2[1 - 2Gm(r)/rc^2]} \qquad (9\text{-}1)$$

where $m(r)$ is the mass within radius r, and $P(r)$, $\rho(r)$ are the pressure and density, respectively, at r. This equation shows that in the case of general relativity pressures contribute to the effective mass–density, and hence the gradient of pressure, dP/dr, is greater than in the nonrelativistic case. The parameters $P(r)$ and $\rho(r)$ are related by the equation of state, which must be known or estimated for each of the density regimes. In general, equations of state are fairly well known at subnuclear densities, but there is considerable uncertainty at high densities (Canuto 1974, 1975b). The equation of state at high densities determines the upper mass limit for neutron stars. An increase in mass increases the central pressure. At some point, depending on the equation of state, the material is unable to support this pressure and the star collapses to form a black hole. Recent investigations (e.g., Pandharipande *et al.*, 1976) suggest that the harder equations of state (greater pressure at a given density) are more realistic, leading to higher upper mass limits for neutron stars, about 2–2.5 $M_\odot$. By requiring that the equation of state not violate causality, Rhoades and Ruffini (1974) have placed a firm upper limit (within general relativity) of 3.2 $M_\odot$ on the mass of neutron stars.

Curves of gravitational mass (proper mass minus the binding energy) and the corresponding moments of inertia are given for two different equations of state in Figure 9-2. For neutron stars with lower mass, central densities are less than or about 2×10^{14} g cm^{-3}, so the crust for such stars is thick and may in fact extend to the center of the star. Generally, the smaller the mass the larger the stellar radius, so that there is an upturn in the moment of inertia at low masses and a downturn at high masses. The minimum stable mass for a neutron star is about 0.1 $M_\odot$.

There is some observational evidence to rule out the softer equations of state (less pressure at a given density), for which the maximum neutron star mass is $<1.3\ M_\odot$. First, as described in Chapter 4, it is likely that the energy required to power the Crab Nebula, about 4×10^{38} erg s^{-1}, originates from the pulsar. If this energy comes from the loss of rotational energy, the moment of inertia (I) must be at least 8×10^{44} g cm^2. For equations of state much softer than Moszkowski's (Figure 9-2), the maximum moment of inertia would be less than this value. Second, limits of neutron star masses can be derived from observations of X-ray pulsars in binary systems. In particular, the mass range of 1.35 to 1.90 $M_\odot$ calculated for Vela X-1 (3U 0900 – 40) (van Paradijs *et al.*, 1976; and see Table 5-1, p. 82) rules out the softer equations of state.

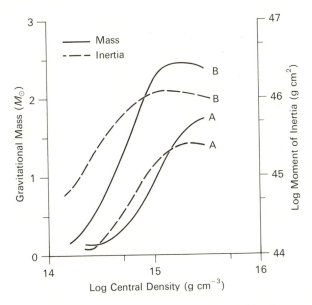

9-2 Gravitational mass (solid lines) and moment of inertia
(dashed lines) of neutron stars as a function of the density
at the center of the star. The A curves are for the equation
of state of Moszkowski (1974), the B curves for the harder
equation of state of Cohen and co-workers. (1970). [After
Canuto, 1975a.]

It is almost certain that the neutron stars we observe as pulsars possess
extremely strong magnetic fields. Calculations based on the rate of loss
of rotational kinetic energy (see the next section) suggest that the surface
magnetic fields of most pulsars are of the order of 10^{12} G. Ruderman and
Sutherland (1973) suggest that turbulent convection in the core of pre-
supernova stars results in the field building up to an equipartition value
of $\sim 3 \times 10^9$ G. Conservation of flux during the collapse to a neutron
star then results in a surface field strength $\sim 4 \times 10^{12}$ G. The field probably has significant multipole structure within a few stellar radii of
the surface, but further out in the magnetosphere it is usually assumed to
be dipolar. There is also likely to be a huge toroidal field interior to the
neutron star. The magnitude of this field is unknown but is probably at
least as strong as the exterior poloidal field (Ruderman, 1972).

Several workers have suggested that decay of the poloidal magnetic
field occurs on a time scale of 10^6 to 10^7 years. The time constant for
field decay is given by

$$\tau_B = 4\sigma R^2 / [\pi(m + 1)c^2], \qquad (9\text{-}2)$$

where σ is the conductivity, R is the stellar radius, and m is an integer representing the order of a multipole field (see for example Chanmugan, 1973). For the lowest order (dipole) component, $m = 0$. Because of proton degeneracy in the interior of the star, the time constant for decay of the interior field is greater than the age of the universe. In the solid crust, however there is no superconductivity, so the exterior field, especially its multiple components, may decay more rapidly. Ewart, Guyer, and Greenstein (1975) find that for stars with very low mass ($\lesssim 0.1\ M_\odot$) and high surface temperatures, significant field decay occurs on a time scale of 10^7 years. For more realistic models, i.e., stars with higher mass and lower surface temperatures, significant decay is unlikely unless the stellar surface is extremely impure or contains large numbers of dislocations.

The strong magnetic fields effectively couple the pulsar magnetosphere to the ions of the crust and charged components in the core of the neutron star. Changes in rotation rate are transferred between these components by Alfvén waves in times of the order of minutes. However, the superfluid neutrons are very weakly coupled to the crust and charged components. Superfluids do not rotate as rigid bodies, but instead form a series of vortex lines parallel to the rotation axis. The number of these vortex lines per unit area (in the equatorial plane) is

$$n_v \approx 2\Omega m_n / \pi \hbar \qquad (9\text{-}3)$$

where m_n is the mass of a neutron and $\hbar$ is Planck's constant (Ruderman, 1976a). For the Crab pulsar, $n_v \approx 2 \times 10^5\ \mathrm{cm}^{-2}$. Each of the vortex lines has a central core of normal neutrons; the only way in which the super-fluid neutrons can change their bulk rotation rate is by interaction of the normal neutrons in the vortex central cores with other components of the star. One possible interaction mechanism is scattering of electrons by the normal neutrons. The radius of a normal core is inversely proportional to the superfluid energy gap, $\mathcal{E}$. When $\mathcal{E}$ is large (for $\rho \approx 2 \times 10^{13}\ \mathrm{g\ cm}^{-3}$, $\mathcal{E} \approx 2$ MeV), the core radius is about 10^{-12} cm, so for the Crab pulsar only about 10^{-18} of the neutrons are normal. This increases the effective interaction time of the neutrons and electrons by a factor of about 10^{18}, from 10^{-11} to about 10^7 s. Feibelman (1971) has shown that this interaction time is proportional to $\exp(\mathcal{E}^2/\varepsilon_f kT)$, where ε_f is the neutron Fermi energy and T is the interior temperature. The strong dependence on $\mathcal{E}$, ε_f, and T allows considerable variation of the computed interaction time; in particular, for older pulsars the temperature would be less and hence the interaction times longer. Feibelman finds that for the Crab pulsar times of the order of a few days are reasonable, whereas for the Vela pulsar times of a year or more could be expected.

Coupling between normal and superfluid components may also occur at the boundaries of an interior zone of normal neutrons. Such a zone will exist if the superfluid energy gap $\mathscr{E}$ becomes less than kT at some radius within the star and is most likely where $\rho \approx 2 \times 10^{14}$ gm cm^{-3} (see Figure 9-1). Ruderman (1976a) suggests that strong interactions may occur where the normal zone merges with the vortex cores, with effective interaction times as short as seconds. Superfluid and charged components would thus spin down together, in which case the density of vortex lines would decrease with time. There is nothing in the interior superfluid zone to prevent this, but in the outer crust the vortex cores may "pin" themselves to crust nuclei. This would prevent spindown of the superfluid material unless the forces became sufficiently great to "unpin" the core or crack the crust. The responses of neutron stars to changes in rotation rate will be discussed further in this chapter in relation to pulsar period measurements (p. 188ff.).

When formed, neutron stars are extremely hot. For the first 10^3 to 10^4 years cooling is by emission of neutrino–antineutrino pairs. At temperatures in excess of 10^{10} K cooling is very rapid, so both solidification of the crust and the onset of superfluidity in the core should occur within about an hour of formation. After 10^3 to 10^4 years cooling is by emission of X-ray photons. Tsuruta and co-workers (1972) have pointed out that the strong surface magnetic fields reduce the photon opacity of the atmosphere and that when the interior becomes superfluid, heat capacities are much reduced. Consequently, cooling by X-ray emission is expected to be relatively rapid, and if there were no other energy inputs, surface temperatures of 10^4 to 10^5 K would be expected after $\sim 10^6$ years. However, frictional interactions causing slowdown of the superfluid neutrons do occur, resulting in thermal dissipation of rotational energy. By including the effects of this heating process, Greenstein (1975) finds that the surface temperatures of older pulsars should be about 6×10^5 K and that they should cool only very slowly. At this temperature the thermal radiation from nearby pulsars should be visible at optical wavelengths ($m_v \approx 22$), especially if they are of low mass and hence large radius. For the Crab pulsar, which is of course much younger, the surface temperature should be $< 2 \times 10^7$ K. The absence of continuous (nonpulsed) low-energy X-ray emission from the pulsar enabled Wolff and co-workers (1975) to set an upper limit of 4.7×10^6 K on the surface temperature of the Crab pulsar, consistent with the theoretical limit. If the interior neutrons were not superfluid the surface temperature should exceed this limit.

Ruderman (1971) has shown that the very strong magnetic fields greatly modify the characteristics of the crystalline crust. Atoms are compressed perpendicular to the field by a factor of the order of 100 and hence are cylindrical in form. The surface layers are therefore expected to form a dense lattice (10^4–10^5 g cm^{-3}), which is a good conductor along the field-lines but effectively an insulator across them. Ions are very tightly bound in this lattice, with a binding energy $W_B \approx 15$ keV. The electric field necessary to remove these ions from the surface is $E_i \approx W_B/Zel$, where l is the lattice spacing of the ions and Z is their charge; for $l \approx 10^{-9}$ cm, $E_i \approx 10^{12}$ V cm^{-1}. Consequently, even the vacuum electric field (see p. 178) is insufficient to remove ions from the surface. The binding energy of electrons is expected to be at least an order of magnitude smaller than that of ions.

The Pulsar Magnetosphere

The radio pulses we observe from pulsars originate somewhere in the space surrounding the star; furthermore, the electromagnetic processes that extract rotational energy from the star occur there. Consequently, an understanding of the properties of this region is of vital importance to the understanding of pulsars. Because of the enormously strong gravitational fields at the surfaces of pulsars ($\sim 10^8$ times larger than the earth's field), the scale height of a normal atmosphere would be very low—about 1 cm for a temperature of 10^6 K. For this reason the first investigations of the pulsar magnetosphere considered essentially vacuum conditions, and we shall describe models based on this premise in the first part of this section.

Prior to the discovery of pulsars, Pacini (1967) pointed out that a rapidly rotating magnetized neutron star would radiate significant amounts of energy in the form of electromagnetic waves—magnetic-dipole radiation—at the rotation frequency. The radiation reaction torque transmitted to the star by the magnetic field is

$$N = -\frac{2(m \sin \alpha)^2}{3c^3} \Omega^3, \tag{9-4}$$

where m is the magnetic-dipole moment and α is the angle between the magnetic and rotation axes. Pacini (1968) and Ostriker and Gunn (1969) recognized that this torque could account for the observed secular increase in pulsar periods. In this case, since $N \propto \Omega^3$, the braking index n (see p. 110) is three. Since the magnetic moment is of the order of $B_0 R^3$, where B_0 is the surface field strength and R is the radius of the star, the magnitude of the surface magnetic field can be estimated from the observed period derivative (with $\sin \alpha$ assumed equal to one), i.e.,

$$B_0 \approx \left(\frac{3Ic^3 P\dot{P}}{8\pi^2 R^6}\right)^{1/2}, \tag{9-5}$$

where I is the moment of inertia of the rotating system. For $I = 10^{45}$ g cm^2, $R = 10^6$ cm, and P in seconds, we have

$$B_0 \approx 3.2 \times 10^{19} \, (P\dot{P})^{1/2} \text{ (gaussian units).} \tag{9-6}$$

Values of B_0 derived from this equation range between about 2×10^{10} and 2×10^{13} G, with fields of about 10^{12} G being typical. The lowest computed surface field is for the binary pulsar PSR 1913+16. It is possible that this pulsar has a different evolutionary history than most, and that its magnetic field is weaker and/or its moment of inertia larger than for other pulsars. Over the range of likely neutron star masses, the quantity I/R^6 may vary by as much as seven orders of magnitude (Greenstein, 1972), compared to a range of five for $P\dot{P}$. It is therefore possible that all neutron stars have the same surface field strength and that the observed range of $P\dot{P}$ arises entirely from variations in stellar mass and hence moment of inertia and radius.

As described in Chapter 4, it is thought that the relativistic electrons in the Crab Nebula are accelerated by the pulsar. Ostriker and Gunn (1969) have shown that the magnetic-dipole model provides an efficient mechanism for accelerating particles. Because the wave field of the magnetic-dipole radiation is extremely strong and of relatively low frequency, particles can be accelerated to almost the wave velocity in less than one cycle and subsequently remain at essentially constant phase in the wave. By means of this "surfing" process, particles can be accelerated to energies as high as 10^{13} eV.

Observational data discussed in Chapter 8 (p. 161) show that pulsars commonly have large spatial velocities, often greater than 100 km s^{-1}. Harrison and Tademaru (1975) have shown that velocities of this order could be produced by an asymmetric radiation-reaction force if the magnetic dipole is offset from the spin axis of the neutron star. If the dipole axis is displaced by a distance s from the spin axis, then the net reaction force is given by

$$F = \frac{2}{15} \left(\frac{\Omega}{c} \right)^5 m_z m_\phi s \tag{9-7}$$

where m_z is the component of the magnetic moment parallel to the spin axis and m_ϕ the component in the azimuthal direction. Because of the strong Ω-dependence, this force is significant only if the star is born with a high rotational velocity ($\Omega \gtrsim 10^4$ rad s^{-1}) and then only for a short time, of the order of one year. However, in this time the pulsar may be accelerated to a high linear velocity, possibly as great as 1000 km s^{-1}.

Rotation of the neutron star and its associated magnetic field produces strong electric fields in the space surrounding the star. Goldreich and

Julian (1969) were the first to point out that because of these fields the region surrounding the star cannot be a vacuum but must contain a substantial space charge. Their argument is as follows: the conductivity of neutron star material is extremely high and may be assumed infinite. Therefore, within the star (provided particle inertia is neglected)

$$\mathbf{E} + \frac{1}{c}(\mathbf{\Omega} \times \mathbf{r}) \times \mathbf{B} = 0, \tag{9-8}$$

where $\mathbf{E}$ and $\mathbf{B}$ are the electric and magnetic fields and $\mathbf{\Omega}$ is the vector angular velocity of the star. The presence of this field requires a redistribution of the charge within the star such that

$$\rho_e = \frac{1}{4\pi}\,\mathbf{V}\cdot\mathbf{E} = -\frac{1}{2\pi c}\,\mathbf{\Omega}\cdot\mathbf{B}. \tag{9-9}$$

The corresponding charge number density is $n_e = 7 \times 10^{-2}B_zP^{-1}$ cm^{-3}, where B_z is the component of $\mathbf{B}$ parallel to $\mathbf{\Omega}$.

In the special case of an aligned dipole field, i.e., one in which the magnetic and rotation axes are parallel, the vacuum magnetospheric fields surrounding the star are static. Laplace's equation, together with boundary conditions at the stellar surface $(r = R)$, gives a quadrupolar electrostatic potential

$$\Phi = -\frac{B_0\Omega R^5}{6cr^3}(3\cos^2\theta - 1), \tag{9-10}$$

where θ is the angle from the rotation axis. The electric field corresponding to this potential has nonzero $\mathbf{E}\cdot\mathbf{B}$, the value at the stellar surface being

$$(\mathbf{E}\cdot\mathbf{B})_R = -\frac{\Omega R}{c}B_0^2\cos^3\theta. \tag{9-11}$$

Consequently, the magnitude of electric field parallel to the magnetic field at the surface is

$$E_{\parallel} \approx \frac{\Omega R}{c}B_0 \approx 6 \times 10^{10}P^{-1} \quad (\text{V cm}^{-1}) \tag{9-12}$$

for $B_0 = 10^{12}$ G and P in seconds. Fields of this magnitude give both electrons and ions an acceleration exceeding that of gravity by many orders of magnitude. The normal factors determining the scale height are therefore completely dominated by electromagnetic effects, and, provided the surface binding energies are not too large, charge will flow from the star to fill the surrounding region.

If particle inertia is neglected, Equations 9-8 and 9-9 also apply to the plasma-filled magnetosphere, and hence the parallel components of the electric field will be zero, i.e.,

$$\mathbf{E} \cdot \mathbf{B} = 0. \tag{9-13}$$

Because of the strong magnetic field, charged particles are forced to co-rotate with the star. However, co-rotation cannot persist beyond the surface where the tangential velocity equals the velocity of light—the so-called *light cylinder*, which has a radius

$$R_{\mathrm{L}} = c/\Omega \approx 5 \times 10^9 P \quad (\text{cm}), \tag{9-14}$$

where P is in seconds. The Goldreich–Julian model of the pulsar magnetosphere is illustrated in Figure 9-3. Two distinct regions are shown: the open field-lines that leave the star near the poles and penetrate the light

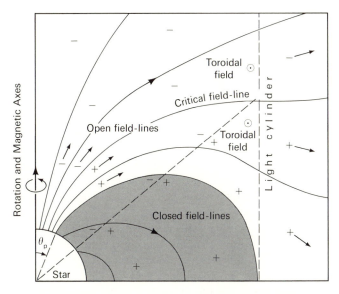

9-3 The magnetosphere of a pulsar with parallel magnetic and rotation axes. Open field-lines cross the light cylinder (where the velocity of co-rotation equals the velocity of light) and are deflected back to form a toroidal field component. The critical field-line divides regions of positive and negative current flow from the star and the plus and minus signs indicate the charge of particular regions of space. The diagonal dashed line is the locus of $B_z = 0$ where the space charge changes sign. [After Goldreich and Julian, 1969.]

cylinder, and the closed field-lines that do not penetrate the light cylinder. In this model the magnetic and rotation axes are parallel; if they were not, the open field-lines would connect to the radiation field. Since $\sin^2 \theta/r$ is constant for dipole field lines, the radius of the polar cap region containing the open field-lines is

$$R_p \approx R \sin \theta_p = R(\Omega R/c)^{1/2}. \tag{9-15}$$

Inasmuch as charged particles are constrained to move along field-lines, at least in the first approximation, they can escape from the star only along the open lines. Given parallel magnetic and rotational axes, as in Figure 9-3, the potential at the base of the field-lines near the axis will be negative with respect to the stellar environment, so electrons will stream from the star along these lines. At some critical field-line, the potential at the stellar surface will be equal to the exterior potential; in the annular region between the critical field-line and the last open line at θ_p it will be positive, so protons (ions) will stream from this region. The location of the critical field-line is determined by the condition that there be no net charge flow from the star.

From Equations 9-10 and 9-15 the potential difference between the center and edge of the polar cap is

$$\Delta\Phi \approx \frac{1}{2} \left(\frac{\Omega R}{c} \right)^2 R B_0. \tag{9-16}$$

For $B_0 \approx 10^{12}$ G and P in seconds, $\Delta\Phi \approx 6 \times 10^{12} P^{-2}$ V. Although the field-lines are equipotential near the star (at least in the present approximation), most of this potential must become available either near or beyond the light cylinder to accelerate charges along the open field-lines. Consequently, particle energies of $\sim 6 \times 10^{12} P^{-2}$ eV are expected, comparable to those expected from acceleration by magnetic-dipole radiation in the vacuum model.

Charge flow along the open field-lines generates a toroidal magnetic field (see Figure 9-3) whose maximum occurs at the critical field-line. The toroidal- and poloidal-field components are expected to be comparable at the light cylinder, so the open field-lines are bent back to cross the light cylinder at an azimuthal angle of about a radian. The corresponding torque on the star may be estimated by integrating the Maxwell stress tensor over a surface surrounding the star. In the approximation that the field-lines within the light cylinder are approximately dipolar, the torque is given by

$$N = -\frac{K}{8c^3} (B_0 R^3)^2 \Omega^3, \tag{9-17}$$

where K is a constant of order unity depending on the detailed field structure. This torque is similar in magnitude to that obtained in the vacuum approximation (Eqn. 9-4) and has the same dependence on Ω, so again the braking index is 3. The braking torque can also be derived by considering the current flow under the polar cap from the electron lines to the proton lines. Because of the similarity of equations 9-4 and 9-17, surface field strengths derived from $P\dot{P}$ on the basis of the axisymmetric model are essentially the same as those obtained for the vacuum model.

The work of Goldreich and Julian considers a simplified system (an axisymmetric rotator), whereas in fact pulsars must possess a non-axisymmetric field (an oblique rotator) in order to produce a periodic signal. Furthermore, Goldreich and Julian do not give a self-consistent description of the currents and fields surrounding the star. Their model involves flow of charge of one sign through a region of space with charge of the opposite sign, a situation unlikely to exist in a real pulsar.

As a first step toward a more realistic model a number of workers have investigated self-consistent solutions that include flow of "massless" particles along the open field-lines. An equation describing the field structure for a steady-state axisymmetric system has been obtained independently by Scharlemann and Wagoner (1973), Michel (1973a), Julian (1973), and Endean (1974). In the force-free approximation, i.e., negligible inertia, the equations of motion (cf. Eqn. 9-8) are

$$\mathbf{E} = -\boldsymbol{\beta}_\pm \times \mathbf{B}, \tag{9-18}$$

and from Maxwell's equations

$$\mathbf{V} \times \mathbf{B} = 4\pi e(Zn_+\boldsymbol{\beta}_+ - n_-\boldsymbol{\beta}_-)$$
$$\mathbf{V} \cdot \mathbf{E} = 4\pi e(Zn_+ - n_-), \tag{9-19}$$

where $\boldsymbol{\beta}_\pm = \mathbf{v}_\pm/c$ for electrons $(-)$ of density n_- and ions $(+)$ of density n_+ and charge Z. From Equations 9-18 and 9-19 we have the equation

$$(\mathbf{V} \cdot \mathbf{E})\mathbf{E} + (\mathbf{V} \times \mathbf{B}) \times \mathbf{B} = 0. \tag{9-20}$$

Solution of this equation with the appropriate boundary conditions gives the self-consistent description of the fields and currents within the approximation of zero inertia. From symmetry about the equatorial plane, the azimuthal component B_ϕ is zero in the closed field-line region; no charge streaming occurs and charge co-rotation is strict on the closed field-lines.

The solution of Equation 9-20 is difficult for realistic field configurations. Michel (1973a) has shown that, for a magnetic monopole, field-lines that remain radial in the meridian plane and form Archimedean spirals in the equatorial plane are an exact solution of the equation. In this case, the outflow velocity is everywhere c and there is no co-rotation. This is similar to the stellar wind case in which the flow energy ($\frac{1}{2}\rho v^2$) exceeds the magnetic energy density ($B^2/8\pi$) by a large factor (e.g., Michel, 1969). In these solutions the radial magnetic-field strength changes as r^{-2} and the torque is proportional to Ω. Consequently, the braking index for such systems is one.

In all of these models the flow is charge-separated—that is, charge of only one sign flows along a given field-line. To avoid the problem of charge flow through regions of space with charge of opposite sign, Michel (1975) has argued that over most of the polar cap the flowing charges have the same sign as the space charge and that the return current is confined to a thin sheath surrounding the open field-lines. In analogy with the earth's field, Michel refers to this sheath as an "auroral zone" on the pulsar.

In the self-consistent analyses discussed above, the effects of particle inertia have been neglected as a first approximation, so $\mathbf{E} \cdot \mathbf{B} = 0$ everywhere in the magnetosphere. It is, however, the parallel electric fields that accelerate charges along the open field-lines, so an estimate of their magnitude is important in determining the maximum energy that such particles may have. Scharlemann (1974) has included inertial effects as a perturbation on the earlier self-consistent analysis, and by numerical integration finds that the parallel component of the electric field is very small; ($E_{\parallel}/E_{\perp} \lesssim 10^{-10}$) so $\gamma_{\pm} = (1 - \beta_{\pm}^2)^{-1/2}$ increases only slowly within the light cylinder. For the case in which the charges are separated, values of $\gamma_{-} \gtrsim 10$ are obtained only close to the light cylinder, and, for the case in which the charge separation is small, only beyond the light cylinder. In a similar analysis Henriksen and Norton (1975a) considered particle acceleration in field structures obtained from the "massless" approximation and were able to obtain analytic solutions to the equations. The results generally confirmed those of Scharlemann (1974) and showed that the acceleration—a "sling" process with the particles constrained to move along the rotating field-lines—was only significant if the streaming current, and hence the bending-back of the field-lines, was small. For this process particle energies are given by

$$\gamma = \gamma_i \gamma_\phi^2 \tag{9-21}$$

where γ_i is the Lorentz factor of the injected particles and γ_ϕ is the co-rotation Lorentz factor defined by

$$\gamma_\phi = (1 - \Omega^2 c^{-2} r^2 \sin^2 \theta)^{-1/2}. \tag{9-22}$$

A second step toward a realistic pulsar model is to relax the requirement of aligned magnetic and rotational axes. Mestel (1971) and Cohen and Toton (1971) were first to show that the conclusions of Goldreich and Julian remain valid for a perpendicular rotator ($\Omega \cdot \mathbf{m} = 0$): charges must flow from the surface of the star to fill the magnetosphere. The form of the field in the force-free (massless particle) approximations and its transition to the wave field outside the light cylinder (i.e., the magnetic-dipole radiation) has been discussed by Endean (1972, 1974), Mestel (1973), and more recently by Henriksen and Norton (1975b). In closed field-line regions where the streaming current is zero, Henriksen and Norton (1975b) found that the field-lines form cusp-like neutral points near the light cylinder, as illustrated in Figure 9-4. In neutron stars with parallel axes, Michel (1973b) found neutral cusps near the light cylinder in the equatorial plane; in those with perpendicular axes, the neutral points form a circle on the surface of the light cylinder, centered on the magnetic axis. It is expected that charge accumulates at these neutral points and is ejected across the light cylinder; the observed radio pulses may originate in such regions where both charge densities and field curvature are high. The form of the open field-line region is modified by the streaming currents as in the axisymmetric case. Henriksen and Norton (1975b) obtain solutions to the force-free equations that relate the field structures in the zone near the star, the light-cylinder region,

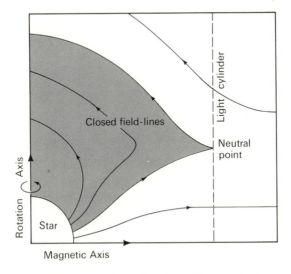

9-4 The magnetosphere of a pulsar with perpendicular magnetic and rotation axes. [After Henriksen and Norton 1975b.]

and the wave zone far from the star. Consistent with earlier work, they find two regions of particle acceleration: the "sling" region inside the light cylinder, and the "surfing" region in the wave zone.

In these force-free models there is essentially no component of the electric field parallel to the magnetic field and hence there is no electrostatic acceleration of particles along the magnetic field lines. However, such parallel electric fields do exist in real pulsar magnetospheres and are probably most significant near the surface of the star, where they accelerate particles injected into the magnetosphere. Sturrock (1971) has considered a model in which essentially all of the potential between the center and edge of the polar cap (Eqn. 9-16) is available to accelerate charges near the surface. Since this potential is $\gtrsim 10^{12}$ V, electrons are accelerated to relativistic factors $\gamma \gtrsim 10^7$. Because of the strong magnetic fields in the polar cap region, the lifetime against synchrotron radiation (Eqn. 4-5, p. 64) is extremely short, so particles move parallel to the field-lines with essentially zero pitch angle. However, if the field-lines are curved, the particles radiate so-called "curvature" radiation in their direction of motion. The characteristic frequency for this radiation, corresponding to the synchrotron relation (Eqn. 4-1, p. 63), is:

$$\omega_c \approx \frac{3}{2}\gamma^3 c/\rho_c, \tag{9-23}$$

where ρ_c is the radius of curvature of the field-line. Thus, for $\gamma \approx 10^7$ and $\rho_c \approx 10^8$ cm, the characteristic frequency is $\omega_c \approx 5 \times 10^{23}$ s^{-1}, i.e., the radiation would be in the form of γ-rays with energy $\varepsilon_\gamma \approx 10^9$ eV. A central feature of Sturrock's model is that γ-rays of this energy moving in a magnetic field of $\sim 10^{12}$ G produce electron–positron pairs. The criterion for pair production is

$$\varepsilon_\gamma B_\perp \geq 4 \times 10^{18} \quad \text{(eV G)} \tag{9-24}$$

where $B_\perp$ is the field component transverse to the photon path; photons moving at a relatively small angle to the field should therefore pair-produce. If the electric field is sufficiently strong, the secondary particles will themselves be accelerated and radiate γ-rays, which in turn can produce further electron–positron pairs. Consequently, a cascade will occur; Sturrock computes that for the Crab pulsar more than 10^7 electron–positron pairs are produced for each primary electron, resulting in a total particle input to the nebula in excess of 10^{40} s^{-1}.

Unfortunately, the central assumption in Sturrock's theory, that the full homopolar potential is dropped along the field-lines near the star, may not be correct. Tademaru (1974) has shown that, taking into consideration the potential variation across the surface of the polar cap region, accelerating potentials parallel to the field may be much smaller than the value assumed by Sturrock (1971). By considering the effects of particle inertia on the distribution of space charge near the star, Michel (1974) has calculated the magnitude of the parallel electric field and hence the energy of the injected electrons. For a pulsar with a period of one second he obtains parallel electric fields of about 5×10^5 V cm^{-1} in the vicinity of the polar cap. The size of the polar cap of a one-second pulsar is about 10^4 cm, so the corresponding electron energies are about 5×10^9 eV, or $\gamma \approx 10^4$. The relation obtained by Michel for the final electron energy is

$$\varepsilon = (8emR^3 B_0)^{1/2}\Omega = 7.5 \times 10^9 P^{-1} \quad \text{(eV)} \qquad (9\text{-}25)$$

so energies are higher for shorter period pulsars. Henrikson and Norton (1976) have investigated this problem on the assumption that fields near the star have a strong multipole component. They find that if the scale, l, of fluctuations in field direction on the surface of the star is small compared to the stellar radius, there exists a zone around the star of thickness comparable to l where parallel electric fields are strong and particle acceleration takes place. They refer to this zone as the *electrosphere* since it is a region of electrostatic acceleration. Because radial magnetic field components of both signs exist in the electrosphere, charges of both signs become accelerated and give rise to an approximately neutral plasma in the magnetosphere. The Lorentz factor for particles injected into the magnetosphere is given by

$$\gamma_i \approx \left(\frac{2el B_0}{mc^2}\right)^{1/2}. \qquad (9\text{-}26)$$

For $l \approx 1$ km and $B_0 \approx 10^{12}$ G, then $\gamma_i \approx 10^7$ for electrons, larger than the values found by Michel (1974) and comparable with those in the model of Sturrock (1971).

All of the above analyses have assumed that both electrons and ions can flow freely from the neutron star surface; however, as discussed earlier (p. 176), it is possible that ions will not be ejected from the surface. Ruderman and Sutherland (1975) have considered the consequences of this restriction on the magnetosphere of an axisymmetric system. They suggest that vacuum gaps form above the neutron star surface in regions where positive charges normally flow (Figure 9-3). In these gaps $\mathbf{E} \cdot \mathbf{B} \neq 0$ and hence field-lines above them are not forced to co-rotate with the star. In the approximation in which the gap height $h < R_p < R$, Ruderman and Sutherland show that the potential across the gap is

$$\Delta\Phi \approx \frac{\Omega B_0 h^2}{c}. \qquad (9\text{-}27)$$

Because of the outflow of positive charge from the outer regions of the magnetosphere, the gap height grows at a velocity near c, and the potential $\Delta\Phi$ rapidly increases toward the upper limit given by Equation 9-16. When the gap potential reaches about 10^{12} V, corresponding to a height of about 5×10^3 cm for a pulsar with a period of about one second, it is discharged by the formation of an electron–positron cascade similar to that discussed by Sturrock. When a pair is produced within the gap, the electron is accelerated toward the surface and the positron toward the outer magnetosphere, each attaining energies of $\gamma_p \approx 2 \times 10^6$. If there is considerable multipole structure near the surface of the star, then the radius of curvature ρ_c is much less than the dipole value $(rc/\Omega)^{1/2}$. For $\rho_c \approx 10^6$ cm these particles will generate curvature radiation at a frequency $\omega_c \approx 5 \times 10^{23}$ s^{-1}, so, as in Sturrock's model, an electron–positron cascade will occur. Beyond the gap, where $\mathbf{E} \cdot \mathbf{B} = 0$, Ruderman and Sutherland predict a streaming electron–positron plasma with $\gamma_s \approx 10^3$. From Equation 9-9 the density of primary positrons at radius r is

$$n_p \approx \frac{\Omega B_0}{2\pi ec}\left(\frac{R}{r}\right)^3 \approx 5 \times 10^{10}\left(\frac{R}{r}\right)^3 \quad (\text{cm}^{-3}), \qquad (9\text{-}28)$$

so the density of secondary particles is

$$n_s \approx n_p\gamma_p\gamma_s^{-1} \approx 5 \times 10^{13}\left(\frac{R}{r}\right)^3 \quad (\text{cm}^{-3}). \qquad (9\text{-}29)$$

In these equations the factor r^{-3} arises because of divergence of the field-lines away from the star, and the numerical factors are appropriate for a one-second pulsar. A similar "outer gap" may form in the region where the diagonal dashed line in Figure 9-3 crosses the open field-lines (Cheng *et al.*, 1976). Inflow of negative charges from this gap (rather than outflow of positive charges from the star) would avoid the problem of charges streaming through regions of opposite sign.

In the models of Henriksen and Norton (1976) and Ruderman and Sutherland (1975) the plasma streaming from the open field-lines is nearly neutral, with density much higher than the minimum charge-separated density (Eqns. 9-9 and 9-28). Okamoto (1974) has pointed out that, apart from Michel's (1973a) monopole solution, there can be no self-consistent axisymmetric solution with complete charge separation in the open field-line region if all field-lines have the same angular velocity. This implies, since co-rotation is enforced by the solid neutron star crust, that the plasma must be normal and not charge-separated. Above the acceleration zone the plasma is almost certain to be highly turbulent.

For example, in the pair-production models primary positrons flow through the secondary plasma, generating fluctuations by means of the two-stream instability (Hinata, 1976). It is also probable that densities are much greater than the minimum charge-separated value in the closed field-line region. In an analysis of the properties of this region Henriksen and Rayburn (1974) conclude that particle densities near the star are about 10^{19} cm^{-3}, and that the plasma is turbulent and hot (to provide support against gravity), with energies of up to about 10^8 eV. Similar conclusions have been reached by Ichimaru (1970) and Kaplan and Tsytovich (1973a).

Many models for the pulse emission process have the pulses originating in the closed field-line region near the light cylinder. It is therefore important to determine how close to the light cylinder the zone of co-rotation extends. Henriksen and Rayburn (1974) point out that co-rotation cannot extend beyond the point where the co-rotation velocity exceeds the Alfvén velocity,

$$v_A = cB(4\pi n\varepsilon)^{-1/2}, \tag{9-30}$$

where ε is the particle energy. (Near the star, $v_A \gg c$.) In their model for the closed field-line region, this condition implies a limit

$$\gamma_\phi < 2^{1/2}, \tag{9-31}$$

where γ_ϕ is the co-rotation Lorentz factor (Eqn. 9-22).

From an analysis of the self-consistent field equation (Eqn. 9-20), Hinata and Jackson (1974) place a more severe limit on the size of the co-rotating zone. If only toroidal currents are permitted, unphysical field structures would occur outside a zone of radius $R_c \approx (R/R_L)^{3/5} R_L$. Consequently, poloidal currents must be permitted, and therefore the field-lines must be open. The boundary of the co-rotating zone is then at about $R_c \approx 6 \times 10^{-3} P^{-3/5} R_L$ and the polar angle of the open field-lines is given by $\sin \theta_p \approx (\Omega R/c)^{1/5}$, much larger than the dipole value (Eqn. 9-15). The braking index for this field structure is 2.2, rather less than the observed value of 2.515 for the Crab pulsar (Chapter 6, p. 112). Roberts and Sturrock (1972a, 1973) obtained a similar braking index by postulating that material collects at the "force-balance" radius, the radius where the orbital period is equal to the pulsar period. Within this radius the field would be approximately dipolar, whereas beyond it would be dominated by inertial effects and so vary as r^{-2} (cf. the stellar wind solution). However, in the force-free models discussed above, material would not be allowed to collect at the force-balance radius, but would be accelerated out toward the light cylinder.

Factors Affecting Pulsar Periods

In Chapter 6 we described the two components of the observed variations of pulsar periods: the regular secular increase and the irregular and unpredictable fluctuations. As described above, the regular increase in pulsar periods is attributed to loss of energy and angular momentum by electromagnetic processes. If the pulsar magnetic field is stable and dipolar in form then the braking index is three. Factors that can change the braking index from this "canonical" value are summarized in Table 9-1.

Ostriker and Gunn (1969) suggest that gravitational radiation may be important in the early stages of a pulsar's life, particularly if the initial rotation frequency is high. If this were so, then the true age of the pulsar would be less than the characteristic age for $n = 3$, $\tau = \frac{1}{2}P/\dot{P}$. For example, gravitational radiation would have to be the principal energy loss process for about the first 80 years to account for the difference between the true age of the Crab pulsar (922 years in 1976) and the characteristic age (1,240 years). The fact that the observed braking index is less than three shows that gravitational radiation is not dominant now.

The low observed value of the braking index for the Crab pulsar also shows that radiation by higher-order multipole magnetic fields is an unimportant contribution to the total energy loss. An approximation to the rate of energy loss is given by the product of the energy density of fields at the light cylinder, the effective area of the light cylinder, and the velocity of light, that is

$$\dot{W} \approx (B_L^2/8\pi)(4\pi R_L^2)c, \tag{9-32}$$

TABLE 9-1

Factors affecting the braking index

	Effect
Multipole electromagnetic radiation	$n \geq 5$
Gravitational quadrupole radiation	$n = 5$
Alignment of dipole field	$n > 3$
Decay of magnetic field	$n > 3$
Radial deformation of field-lines	$1 \leq n \leq 3$
Counteralignment of dipole field	$n < 3$
Relaxation of equilibrium-form of neutron star	$n < 3$
Transverse velocity	$n < 3$

where B_L is the magnetic field strength at the light cylinder. If $B_L = B_0(R/R_L)^p$, where R is the stellar radius, one obtains

$$\dot{W} \approx \tfrac{1}{2} c^{3-2p} B_0^2 R^{2p} \Omega^{2p-2} \tag{9-33}$$

and $n = 2p - 3$. For a dipolar field, $p = 3$ and $n = 3$, as described above. Outward radial deformation of the field-lines reduces p and hence n, with the limiting value being the stellar wind solution, where $p = 2$ and hence $n = 1$. Such deformation is the most likely reason for observed braking indexes less than 3. Magnetospheric models with this type of field structure were described earlier (p. 187).

Several authors have considered the question of a possible secular change in orientation of the magnetic field relative to the rotation axis, and various conclusions have been reached. In a model in which the neutron star is assumed to be a perfectly conducting sphere and magnetic-dipole radiation is the principal energy loss process, radiation-reaction torques cause the dipole axis to align with the rotation axis over a time comparable to the pulsar lifetime. The resulting braking index, $n = 3 + 2 \cot^2 \alpha$, where α is the angle between the dipole and rotation axes, is thus always greater than three and increases with time. If the vacuum condition were relaxed, n would not tend to infinity as $\alpha \to 0$, since currents would continue to brake the rotation. The conclusions are also modified if the effects of crust rigidity and nonsphericity of the neutron star are included. The two main effects likely to cause nonsphericity are centrifugal and magnetic stresses. If only centrifugal stresses are considered, the equilibrium shape of a rotating star is an oblate spheroid with moments of inertia

$$I_1 = I_2 = I_0 \left(1 - \frac{5}{16\pi} \frac{\Omega^2}{G\rho} \right) \tag{9-34}$$

$$I_3 = I_0 \left(1 + \frac{5}{8\pi} \frac{\Omega^2}{G\rho} \right) \tag{9-35}$$

about the equatorial and symmetry axes, respectively, where $I_0 = \tfrac{2}{5} MR^2$ and $M = \tfrac{4}{3} \pi R^3 \rho$. In a star with significant rigidity the rotation and symmetry axes need not coincide; the rotation axis then undergoes a free precession about the symmetry axis. If the star were perfectly rigid the precession frequency would be

$$\Omega_p = [(I_3 - I_1)/I_1] \Omega \cos \Theta, \tag{9-36}$$

where Θ is the angle between the symmetry axis and the instantaneous rotation axis. In a star of lesser rigidity most of the deformation follows the instantaneous rotation axis, so the precessional frequency is smaller than that given by Equation 9-36. Goldreich (1970) has shown that if the angle between the magnetic dipole moment and the symmetry axis (χ) is less than $55°$, radiation torques tend to damp out the precessional motion. Conversely, if $\chi > 55°$, the precessional angle (Θ) increases. Precessional motion is also damped by frictional dissipation of energy in the crust as the equatorial bulge adjusts to the instantaneous rotation axis. The time scale for damping of the precessional motion by frictional effects depends on the rather poorly known properties of the crust; it is generally believed that damping forces dominate over forces that tend to increase the precessional angle. One would therefore expect the rotation and symmetry axes to coincide and hence α to equal χ. However, if the symmetry axis is determined by the magnetic field, i.e., if the magnetic distortion is greater than other distortions that do not follow the instantaneous rotation axis, then either alignment or counter-alignment ($\alpha \to 90°$) will occur. If the field interior to the neutron star is predominantly poloidal, then $\chi = 0$ and there will be alignment; if it is predominantly toroidal with an axis coincident with that of the external poloidal field, then $\chi = 90°$ and counter-alignment is to be expected. Alignment or counter-alignment can also occur if the magnetic axis rather than the rotation axis moves through the star. Macy (1974) has considered various combinations of parameters in models in which braking is by magnetic-dipole radiation. In counter-aligning models n can be as low as 2 during the period of counter-alignment ($10^5–10^6$ years after formation). When counter-alignment is complete, $n = 3$.

As mentioned earlier (p. 173), significant decay of the poloidal magnetic field may occur during the pulsar lifetime. This would result in a decrease in the energy loss rate similar to that resulting from alignment, and hence an increase in the braking index. Again, at least for the Crab pulsar, it appears that magnetic decay is not significant at the present time. However, it may be significant for older pulsars for which n has not yet been determined.

As a spinning neutron star slows down, its equilibrium axial moment of inertia I_3 (Eqn. 9-35) steadily decreases. Provided the crust rigidity does not prevent it, the actual moment of inertia likewise decreases. Consequently, the rate of slowdown and the braking index are less than they would otherwise be. From Equations 6-17 (p. 112) and 9-35 we find that the perturbation in the braking index is given by

$$\Delta n = -\frac{5n\Omega^2}{4\pi G\rho} \tag{9-37}$$

For $\Omega = 200 \text{ s}^{-1}$ (the Crab pulsar), $n = 3$, and a density of $10^{14} \text{ gm cm}^{-3}$, $\Delta n = -0.0072$, so the effect is small. It does, however, have some importance in the interpretation of period irregularities, as described below.

Shklovskii (1969) has pointed out that for a pulsar moving across the line of sight with velocity v at a distance d there is an effective contribution to the secular slowdown of

$$\Delta\dot{\Omega} \approx -v^2\Omega/cd \qquad (9\text{-}38)$$

because of the continually changing radial velocity. Provided the intrinsic braking index is greater than one, large transverse velocities result in an observed braking index less than the intrinsic value, the perturbation being given by

$$\Delta n = -(n - 1)\,\Delta\dot{\Omega}/\dot{\Omega}. \qquad (9\text{-}39)$$

For PSR 1133+16, which has a transverse velocity of 310 km s^{-1} (Table 8-2, p. 161) and a distance of about 180 pc, the contribution to $\dot{\Omega}$ is -3.0×10^{-16} s^{-2}, or about 1.8 percent of the observed value. Thus, assuming $n = 3$, the observed braking index is reduced by about 3.6 percent.

The phenomena listed in Table 9-1 and discussed above concern the secular or long-term variations in pulsar periods. As described in Chapter 6 (p. 113ff.), period irregularities on much shorter time scales are observed in many pulsars. The most dramatic of these are the abrupt spinups observed in the Crab and Vela pulsars. Shortly after the first Vela spinup Baym and co-workers (1969) proposed that a sudden cracking of the neutron-star crust (a *starquake*) results in a decrease of the moment of inertia and hence an increase in Ω. If we define the oblateness of the star by $I = I_0(1 + \varepsilon)$, then the total energy of the rotating star is

$$W = W_0 + \frac{1}{2} I\Omega^2 + \mathscr{A}\varepsilon^2 + \mathscr{B}(\varepsilon_0 - \varepsilon)^2, \qquad (9\text{-}40)$$

that is, the energy of the nonrotating star, (W_0), plus the rotational kinetic energy ($\frac{1}{2}I\Omega^2$), the gravitational deformation energy ($\mathscr{A}\varepsilon^2$), and the elastic strain energy ($\mathscr{B}(\varepsilon_0 - \varepsilon)^2$). The coefficients are defined as $\mathscr{A} \approx \frac{3}{25}GM^2/R$ and $\mathscr{B} \approx \frac{1}{2}\mu V_c$ (where μ is the mean shear modulus of the crust and V_c is its volume), and ε_0 is a reference oblateness. Minimizing the total energy with $I\Omega$ and ε_0 constant gives to first order

$$\varepsilon = \frac{I_0\Omega^2}{4(\mathscr{A} + \mathscr{B})} + \frac{\mathscr{B}}{\mathscr{A} + \mathscr{B}}\varepsilon_0. \qquad (9\text{-}41)$$

Estimates show that in neutron stars $\mathscr{A} \gg \mathscr{B}$ (e.g., Baym and Pines, 1971) so

$$\varepsilon \approx I_0\Omega^2/4\mathscr{A} + (\mathscr{B}/\mathscr{A})\varepsilon_0. \qquad (9\text{-}42)$$

For $\mathscr{B} = 0$ this reduces to the perfect fluid value (cf. Eqn. 9-35). As the star slows down, the oblateness decreases. However, because of its rigidity, the crust resists deformation and a stress

$$\sigma = \mu(\varepsilon_0 - \varepsilon) \tag{9-43}$$

builds up. At some Ω, the stress exceeds a critical value and the crust cracks, resulting in a decrease in the reference oblateness, $\Delta\varepsilon_0$, and a corresponding decrease in the actual oblateness, $\Delta\varepsilon \approx (\mathscr{B}/\mathscr{A}) \Delta\varepsilon_0$. Since

$$\Delta\varepsilon = \Delta I/I = -\Delta\Omega/\Omega, \tag{9-44}$$

these decreases result in an increase in the rotation rate. The computed value for the Vela spinups is $\Delta\varepsilon \approx 2 \times 10^{-6}$, a rather large fraction of the current equilibrium value of $\varepsilon \approx 5 \Omega^2/(8\pi G\rho) \approx 1.5 \times 10^{-4}$.

As mentioned in Chapter 6, at the time of the first Vela spinup a change $\Delta\dot{\Omega} \approx 10^{-2}\dot{\Omega}$ accompanied the increase in Ω. Baym and co-workers (1969) proposed an explanation for this increase in $|\dot{\Omega}|$ based on the two-component (superfluid and nonsuperfluid) model of the neutron star. Because the external braking torque (N) acts on the charged, nonsuperfluid component (the crust and any charged components in the core), the charged and neutron components slow down according to the relations

$$I_c\dot{\Omega} = -N + (I_c/\tau_r)(\Omega_n - \Omega) \tag{9-45}$$

and

$$I_n\dot{\Omega}_n = -(I_c/\tau_r)(\Omega_n - \Omega), \tag{9-46}$$

where I_c and I_n are the moments of inertia of the charged components and the superfluid neutrons, respectively; Ω_n is the bulk rotation rate of the neutrons (assumed constant throughout the star); and τ_r is a relaxation time describing the frictional coupling between the crust and the neutrons. When the system is in equilibrium, $\dot{\Omega} = \dot{\Omega}_n$ and

$$\Omega_n - \Omega = \frac{I_n}{I_c}\frac{\tau_r}{T}\Omega, \tag{9-47}$$

where $T = -\Omega/\dot{\Omega}$, the characteristic time. Consequently, the inner neutron superfluid rotates more rapidly than the crust. In the Vela pulsar τ_r is thought to be a few years (see p. 174 and below), so for $I_n/I_c \approx 1$, i.e., a moderately massive neutron star, $(\Omega_n - \Omega)/\Omega \approx 10^{-4}$. Older stars are likely to have a lower interior temperature and hence a much larger τ_r; it is possible that in such pulsars Ω_n is many times larger than Ω (Greenstein, 1975).

Immediately after a starquake, Ω is increased and $\Omega_n - \Omega$ is less than its equilibrium value. The net braking torque on the charged components is therefore increased, resulting in an increase in $\dot{\Omega}$ until equilibrium is restored. From Equation 9-45, neglecting small terms,

$$\frac{\Delta\dot{\Omega}}{\dot{\Omega}} = \frac{\Delta\Omega}{\Omega}\frac{T}{\tau_r}\left[1 - \frac{\Delta I_n/I_n}{\Delta I_c/I_c}\right] \tag{9-48}$$

so that if as expected $\Delta I_n/I_n \ll \Delta I_c/I_c$, the fractional increase in $|\dot{\Omega}|$ is about T/τ_r times larger than that in Ω. From the observed values of $\Delta\dot{\Omega}/\dot{\Omega}$ and $\Delta\Omega/\Omega$ for the first Vela spinup (Table 6-1, p. 118), $\tau_r \approx 3 \times 10^{-4}T \approx 6$ years. If the behavior of pulsars is adequately described by this two-component model, the ratio of $\Delta\dot{\Omega}/\dot{\Omega}$ to $\Delta\Omega/\Omega$ should be the same for different spinups in a given pulsar, regardless of the mechanism producing the initial jump (Greenstein, 1976). Any change in this ratio could only result if the internal temperature of the star changed as a result of the spinup or ΔI_n was not zero and not proportional to ΔI_c. For the Vela pulsar, all three observed spinups have had similar parameters (Table 6-1), so the ratio has also been constant. However, the ratio for the 1975 spinup of the Crab pulsar is about a factor of four smaller than that for the 1969 jump. In the models of Greenstein (1976), the changes in $\Delta\dot{\Omega}/\dot{\Omega}$ and $\Delta\Omega/\Omega$ are very rapid in the first few hours after a spinup. Such behavior has not yet been observed but, if present, it would modify the conclusions drawn from Equation 9-48.

From Equations 9-45 and 9-46, the time dependence of Ω after the spinup is

$$\Omega(t) = \Omega_0(t) + \Delta\Omega[1 - Q(1 - e^{-t/\tau_d})] \tag{9-49}$$

where $\Omega_0(t)$ is the value of Ω extrapolated from before the event, Q is a parameter describing the degree to which the frequency relaxes back toward its extrapolated value, and $\tau_d = \tau_r I_n/I$ is the time constant of the decay. Figure 9-5 shows the expected time-dependence of Ω. The parameter Q is a measure of the fraction of the original speedup, $\Delta\Omega$, which decays away. In the context of the simple two-component model Q may be expressed in terms of the changes in Ω and its derivatives at the time of the jump, and is related to the relative moments of inertia of the charged components and the neutrons as follows:

$$Q = \frac{(\Delta\dot{\Omega})^2}{\Delta\ddot{\Omega}\,\Delta\Omega} = \frac{I_n}{I}\left[1 - \frac{\Delta I_n/I_n}{\Delta I_c/I_c}\right] \approx \frac{I_n}{I} \tag{9-50}$$

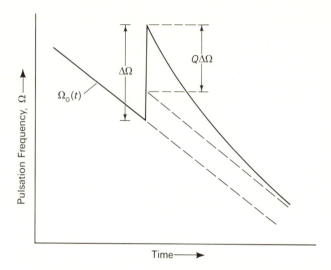

9-5 Time dependence of the pulsar angular frequency, Ω, following a discontinuous speedup. See Equation 9-49.

for small ΔI_n. Therefore,

$$\tau_d \approx Q\tau_r \approx TQ \frac{\Delta\Omega/\Omega}{\Delta\dot\Omega/\dot\Omega}. \qquad (9\text{-}51)$$

For a light neutron star most of the inertia is in the superfluid component, so after relaxation the frequency will be close to the extrapolated value, i.e., Q will be close to one. For a heavy star most of the inertia may be in a charged (perhaps solid) core that is spun up with the crust, so very little relaxation will occur and Q will be close to zero. Some superfluid neutrons will always be present and the minimum Q is thought to be about 0.05. These results suggest that the Crab pulsar is a light neutron star ($M \lesssim 0.5\ M_\odot$) and that the Vela pulsar is heavier ($M \gtrsim 1\ M_\odot$).

The starquake model is, however, unable to account for the relatively short intervals between the Vela spinups. For a second starquake to occur, the stress (Eqn. 9-43) resulting from the change in oblateness as the star slows down must again build up to the critical value; the time required for this, given by Baym and Pines (1971), is

$$t_q = \frac{2\mathscr{A}^2}{\mathscr{B}I_0} \frac{T}{\Omega^2} |\Delta\varepsilon|. \qquad (9\text{-}52)$$

For the Vela pulsar, $t_q \gtrsim 10^3$ years, whereas in fact the Vela events were separated by only 2.5 and 4.0 years. Furthermore, considering that the events occur every few years, the observed speedups represent a rather large fraction (about one percent) of the current equilibrium oblateness. Two possible modifications of the starquake model have been suggested to overcome these problems. If the neutron star possesses a solid core, the principal inertia change may occur as a *corequake* (Pines *et al.*, 1972). Since the shear modulus of the core is expected to be several orders of magnitude greater than that of the crust, its present oblateness could be much larger than the current equilibrium value. Consequently, the relatively large and frequent events observed in the Vela pulsar may possibly be accommodated in the corequake model. The main problems with this model are whether the core is solid (see p. 171) and the amount of energy dissipated in the core. If corequakes of magnitude $\Delta\varepsilon \approx 10^{-6}$ occurred every few years, with each releasing about 10^{45} ergs of strain energy, then the neutron star would be heated to the point where it would become a significant source of X-rays (Pandharipande *et al.*, 1976). Such X-rays are not observed, however. If the core were solid, its precessional frequency

$$\Omega_p = \frac{3\varepsilon_0 \mathscr{B}}{\mathscr{A} + \mathscr{B}} \Omega \qquad (9\text{-}53)$$

would be about three orders of magnitude larger than if it were liquid (i.e., a precessional period of minutes rather than days) (Pines *et al.*, 1972). If the precessional angle were sufficiently large, there could be a detectable variation in the pulse period (see below) and/or the pulse amplitude, but such an effect has not been detected in any pulsar.

A quite different interpretation for the large Vela spinups has been suggested by Anderson and Itoh (1975) and Ruderman (1976a). Interaction between the normal cores of vortex lines and a central zone of normal neutrons would slow the superfluid neutrons with a relatively short time scale (see p. 175). However, if the vortex cores are pinned to crust nuclei, the superfluid would be prevented from slowing down. In regions where the superfluid energy gap, $\mathscr{E}$, is large, the resulting stresses would be likely to crack the crust rather than unpin the vortex lines. This would result in a transfer of angular momentum from the superfluid neutrons to the crust, thus speeding up the crust. Provided a central zone of normal neutrons exists, only a small fraction of the superfluid neutrons would be involved in this momentum transfer and the fractional spinup is expected to be close to the observed value ($\Delta\Omega/\Omega \approx 2 \times 10^{-6}$). Of course, in this model a different mechanism must be found for the post-spinup decay of Ω. Ruderman (1976a) suggests that lattice creep occurs in response to the large stresses placed on the crust by the vortex cores, effectively reducing the observed slowdown rate

$|\dot{\Omega}|$. These stresses would be relieved when the crust cracked, so the observed slowdown rate would increase. Estimates of the magnitude of creep relaxation show that the observed $\Delta\dot{\Omega}/\dot{\Omega} \approx 10^{-2}$ could be achieved.

Other mechanisms proposed to account for the observed discontinuities in rotation frequency include magnetospheric instabilities (Scargle and Pacini, 1971; Roberts and Sturrock, 1972b) and instabilities in the motion of the superfluid neutrons (Greenstein and Cameron, 1969). The apparent association of activity in the nebular wisp features of the Crab Nebula (p. 68) with discontinuities in the Crab pulsar period led to explanations based on the sudden release of particles trapped in the closed field-line region. The maximum plasma moment of inertia that the pulsar magnetic field can sustain is $I_p \approx B_0^2 R^3/6 \, \Omega^2$ (Ruderman, 1972). If all of this plasma were suddenly released (without producing a torque on the star), then $\Delta\Omega/\Omega \approx I_p/I \approx B_0^2 R/M\Omega^2$, leading to $\Delta\Omega/\Omega \lesssim 10^{-6}$ for the Vela pulsar and $\Delta\Omega/\Omega \lesssim 10^{-7}$ for the Crab pulsar. It is therefore possible that the observed Crab events result from this effect, but it is unlikely that those in the Vela pulsar do. Problems with this explanation concern the question of whether or not sufficient mass accumulates in the magnetosphere, and if it does, why it is stored for such long intervals and released in a single large event.

Fluctuations in the rate of transfer of angular momentum from the neutrons to the crust could result if the neutron superfluid were turbulent. However, Ruderman and Sutherland (1974) have shown that the vortex lines in the superfluid tend to remain parallel to the rotation axis, thereby suppressing any tendency to turbulence in the superfluid.

As described in Chapter 6, irregular frequency jumps of either sign with $|\Delta\Omega| \lesssim 10^{-9}$ Hz are observed in several pulsars, especially those with large values of the parameter $\dot{P}P^{-5}$. Pines and Shaham (1972) have attributed these events to *microquakes* in the neutron star crust resulting from stresses induced by a linearly growing precessional angle. Such an increasing precessional motion could occur if the radiation torque were aligned with the plane formed by the symmetry axis of the star and the angular momentum vector, rather than remaining fixed in the star. If the torque moves in this way, stresses resulting from the precessional motion build up on a shorter time scale than those resulting from the variation in the oblateness (ε), and are released by microquakes in which there is both a decrease in the precessional angle (Θ) and a small decrease in ε. The sign of the resulting frequency jump depends on the relative magnitudes of $\Delta\Theta$ and $\Delta\varepsilon$ and may vary. Pines and Shaham estimate that in the Crab pulsar, frequency jumps with $\Delta\Omega/\Omega \approx 10^{-11}$ could occur several times a day if the current precessional angle were about three degrees. This would be sufficient to account for the frequency irregularities observed in the Crab pulsar (Figure 6-7, p. 119). Chau,

Henriksen, and Rayburn (1971) have suggested that if the neutron star were triaxial, some of the observed irregularities could result directly from the precessional motion.

A number of other mechanisms have been proposed to account for the observed random variations in pulsar periods. Differential rotation in the superfluid core, especially following a spinup, would excite various modes of Tkachenko oscillations of the vortex lines (Ruderman, 1972). The resulting variable frictional torque on the neutron star crust could contribute to the observed irregularities. Another possible mechanism suggested by Ruderman is cracking of the neutron star crust induced by magnetic stresses. If the crust solidified before these stresses were relieved, it would have to sustain stresses $\sim B^2/8\pi \approx 10^{23}$ dynes cm^{-2} in order to stabilize the field. In weaker parts of the crust this may not be possible. As mentioned above, vortex cores in the neutron superfluid are likely to be pinned to nuclei in the crust. Anderson and Itoh (1975) suggest that in some pulsars the slowing-down of the superfluid, and hence of the crust, proceeds in an irregular fashion as the vortexes "creep" outwards. Where the vortexes are so strongly pinned that no creep is possible, the superfluid would have a constant angular velocity and no irregularities in the spindown rate of the crust would occur. This may be the case for most pulsars. Finally, there is the possibility that period irregularities result from small changes in the structure and content of the magnetosphere. In the absence of a full understanding of the magnetosphere, this idea has not been explored in any detail.

10

Pulse Emission Mechanisms

One of the least understood aspects of pulsars is the mechanism by which rotational energy of the neutron star is converted into the pulses we observe. Although numerous theoretical models for the emission mechanism have been proposed, no single model has been generally accepted, partly because of the great diversity of the observational data. Some authors have in fact suggested that different pulsars may have different emission mechanisms. However, excepting optical and higher-frequency radiations, the characteristics of the radio emission from the various pulsars are sufficiently similar that a single basic model probably applies to all pulsars. That model has not yet been specified to the satisfaction of most astrophysicists, but it is likely to be described at least partly by the mechanisms proposed so far.

In this chapter we shall first review the basic observational requirements for the pulse emission mechanism. The most important of these are that the radiation be emitted in a relatively narrow beam and that the mechanism be capable of producing an extremely high specific intensity (or brightness temperature) of broadband radiation. Coherent mechanisms capable of producing this high-intensity radiation are then described. The proposed emission mechanisms may conveniently be divided according to their location with respect to the neutron star: processes occurring outside the light cylinder, those in which the emission originates in the closed field-line region of the magnetosphere, and those associated with the open field-line region. These are discussed in turn. Following this, optical, X-ray, and γ-ray emission mechanisms are described. Finally, we discuss models for subpulse drift and the evolution of pulsars toward the point at which they no longer emit significant pulsed radiation.

Observational Requirements

It is now generally accepted that pulsars are rotating neutron stars. Thus, the most basic requirement of any proposed emission mechanism is that it generate a beam of radiation that is fixed in orientation with respect to the neutron star. The beam must have a width in the longitude direction of about 10° (as seen by an outside observer) and this width must remain nearly constant over several decades of frequency. Because integrated profiles are often rather complex in shape and (except for mode changing) are extremely stable, the factors determining the beam shape and longitude must be similarly stable.

The frequent occurrence of "double" integrated profiles shows that radiation beams that are effectively double-lobed are common. For example, in models in which the emission is beamed radially outward along the open field-lines, i.e. *magnetic pole* models, double profiles would be expected if the radiation beam were in the form of a hollow cone. In such a model, if the line of sight and the magnetic axis make about the same angle with the rotation axis, a double profile will be observed, whereas if the line of sight passes near the edge of the radiation beam as the star rotates, a single profile will be observed. In other models the integrated profile is determined by the distribution in longitude of the emitting particles. For example, in some models the distribution of emitting particles around the light cylinder determines the integrated profile.

Interpulses are generally associated with an emission source located on the side of the star opposite from the main source. In magnetic-pole models interpulses are identified with radiation from the opposite pole of a predominantly dipole field and are visible when the magnetic and rotation axes are nearly perpendicular. This model is supported by the similar polarization characteristics of the main pulse and interpulse of the Crab pulsar (Figure 4-8, p. 74) and is also consistent with the observed fraction of pulsars possessing interpulses.

However, several observations suggest that main pulses and interpulses may be more closely related. First, all pulsars having interpulses (with the exception of PSR 0904 + 77, for which the adopted period may be twice the true value) have periods less than 0.6 s. Unless the magnetic axis always tends to align with the rotation axis, interpulses would be expected in pulsars of any period. Second, radio observations of the Crab pulsar have shown that there is a very high correlation between the long-term intensity variations of the main pulse and interpulse (Figure 4-7, p. 73). It is not immediately obvious why this should be so if the main pulse and interpulse originate in widely separated locations. Third, significant emission occurs between the main pulse and interpulse in PSR 0950+08 (Figure 2-2, p. 17), in the optical and X-ray profiles of the Crab pulsar (Figure 4-6, p. 70), and in the optical and γ-ray profiles of the Vela pulsar (Figure 5-1, p. 79). All of these profiles are in fact similar to those for two-component pulsars, such as PSR 0525+21 and PSR 1133+16, except for the much wider separation of the components. Finally, the smaller spacing and symmetric location of the Vela optical pulse components with respect to the γ-ray components strongly suggest beaming from a single polar region. It may be that there is no fundamental difference between pulsars with two-component profiles and those with interpulses. For example, in magnetic-pole models an interpulse would be observed if the half-angle of the cone were close to 90 degrees, making the beam fan-shaped.

The emission process (or processes) must be capable of producing broad-band radiation at both radio and optical frequencies. At radio frequencies individual pulses have bandwidths in excess of 100 MHz; it is likely that in many pulsars the spectrum of individual pulses does not differ significantly from that of the integrated profile. At optical frequencies there is no evidence for fluctuations in the shape or intensity of pulses from the Crab pulsar, so the spectrum of every pulse must be identical with that of the integrated profile.

The emission processes must be able to produce the observed luminosities and brightness temperatures of the radio, optical, and X-ray emission. Observed luminosities (assuming a conical beam) are in the range 10^{25} to 10^{28} erg s^{-1} for most pulsars. If we assume a source area of 10^{15} cm^2, across which the light travel time is 1 ms, the specific intensities corresponding to these luminosities are extremely high (10^4–10^7 ergs cm^{-2} s^{-1} Hz^{-1} sterad^{-1}), and brightness temperatures are in the range 10^{23} to 10^{26} K. Observations of microstructure with characteristic time scales of the order of 100 μs in several pulsars, and also the occasional intense pulses from the Crab pulsar, imply brightness temperatures as high as 10^{30} to 10^{31} K. As mentioned in Chapter 1, these extremely high brightness temperatures cannot be produced by any incoherent radiation mechanism because implausibly high particle energies would be required. Furthermore, even if such high-energy particles were available, they would not radiate in the radio range in which most pulsar emission occurs.

The luminosity of the optical and X-ray emission from the Crab pulsar is very high, about 10^{35} erg s^{-1}, but the implied brightness temperatures are less than or about 10^{11} K, much lower than at radio frequencies. For incoherent radiation of this intensity, particles of energy $\varepsilon \gtrsim kT \approx 10^7$ eV are required. Such energies are similar to those expected in pulsar magnetospheres, so the optical radiation need not be coherent. This is consistent with the observed lack of intensity fluctuations in the optical pulses. As the plasma in the emitting region is likely to be turbulent, the absence of intensity fluctuations also implies that the optical pulses are emitted from a region that is large relative to the scale of the turbulence.

At radio frequencies both individual pulses and integrated profiles often have strong linear polarization. In integrated profiles the position angle often varies smoothly through the profile, with the total change less than or about 180°. This position angle variation is independent of frequency and, like the integrated profile itself, stable over long time intervals. The lack of frequency dependence implies that the change of position angle throughout the pulse is a geometric effect, presumably related to the orientation of magnetic fields in the emission region, and not the result of propagation effects occurring in the pulsar magnetosphere or the interstellar medium. (This will be discussed further in this chapter, p. 225.)

In some models, for example those of Eastlund (1968) and Ferguson (1973), the integrated profile is assumed to represent the beam shape of radiation from a single group of particles. Variation in polarization position angle within the profile is then related to variation in the projected direction of a single vector attached to the rotating star. In other models, for example those of Radhakrishnan and Cooke (1969) and Smith (1970), the radiation beam from a single group of particles is much narrower than the integrated profile (either subpulses or micropulses may be identified with the elementary beam). The integrated profile then represents a spatial distribution of emitting regions, each with its own projected field direction.

A number of observations tend to favor this second interpretation. First, integrated profiles differ greatly from one pulsar to another and are often complex in form, with some pulsars having five separate and identifiable components. It seems unlikely that radiation from a single group of particles would have such a complex beam shape. Second, integrated profiles are extremely stable over long time intervals; the single-vector model would necessitate a similar stability in particle parameters, such as energy and pitch angle. Third, in general, different parts of the integrated profile have different fluctuation characteristics and different spectral indexes, and the polarization behavior of individual subpulses varies across the profile. For example, orthogonally polarized subpulses are usually observed in only part of the profile (p. 51). Finally, the mode-changing phenomenon (p. 30) is difficult to explain if the entire integrated profile is generated by a single group of particles. Where different particles contribute to different parts of the profile, mode-changing could result from a change in their spatial distribution.

A number of correlations exist among the various properties of pulsars. These correlations should be explained by, or at least be not inconsistent with, any satisfactory pulsar model. As was described in Chapter 2 (p. 15), pulsars can be divided into two classes according to their profile shape: Type C pulsars have a complex (usually double) profile, whereas Type S pulsars have a simple, single-peaked profile. Both of these classes include subclasses of pulsars that show evidence of drifting subpulses (Types CD and SD). Many of the correlations are conveniently described in terms of these classes. First, as shown in Figure 10-1, profile shape is dependent on period, with the majority of Type C pulsars having periods

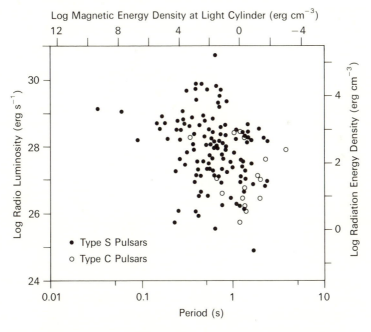

10-1 Radio luminosity of pulsars as a function of period. Luminosities were computed from the observed pulse energy at 400 MHz and assuming a bandwidth of 400 MHz and a conical radiation beam of angular diameter $2\pi W_e/P$, where W_e is the pulse equivalent width. Also shown are the corresponding radiation energy densities at the source, computed assuming a source area of 10^{15} cm^2, and magnetic energy densities at the light cylinder, computed assuming a dipolar field of strength $B_0 = 10^{12}$ G at $R = 10^6$ cm. Effects of co-rotation beaming on the radiation energy densities have not been included.

exceeding one second. The possibility that long-period pulsars have a different evolutionary history from short-period pulsars was mentioned in Chapter 8 (p. 153). Figure 10-1 also shows only a weak correlation between period and luminosity; however, it is notable that most Type C pulsars have rather low luminosities.

Period derivatives tend to be large for Type C pulsars and small for Type D pulsars, especially those in the subclass SD. There is also a correlation between direction of subpulse drift and period derivative, as shown in Figure 10-2. Without exception, pulsars with subpulses drifting pre-

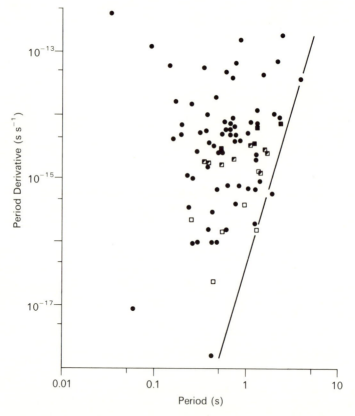

10-2 Period derivative plotted against period for 87 pulsars. The sloping line represents a constant magnetic-field strength (~ 2 G) at the light cylinder, assuming a fixed value (10^{45} g cm^2) for the stellar moment of inertia. Type D pulsars are plotted as squares; the squares are open if the direction of subpulse drift is from the trailing edge to the leading edge of the integrated profile; filled if the drift is in the opposite direction; and half-filled if both directions of drift are seen.

dominantly toward the leading edge of the profile have small period derivatives. Furthermore, these sources are all of type SD. Some pulsars with relatively high values of $\dot{P}$ have occasional, and usually rather irregular, drifting subpulse behavior. For the pulsars with the highest values of $\dot{P}$ the drift direction is in general toward the trailing edge of the profile, whereas for intermediate values of $\dot{P}$ both directions of drifting are observed. As pointed out by Ritchings and Lyne (1975) these correlations make untenable any model in which the character of drifting subpulses is dependent on observer location relative to the rotation or magnetic axis of the star.

The sloping line in Figure 10-2 represents a constant value of the parameter $\dot{P}P^{-5}$. Evidently, pulsed emission cannot be generated if this parameter is less than the value represented by the line (about 5×10^{-17} s^{-5}). This correlation will be discussed in some detail later in the chapter (p. 233).

Coherent Emission

By definition, the specific intensity (I_v) from a coherent emission source containing N particles must satisfy the condition

$$I_v > NI_{v,\,i} \tag{10-1}$$

where $I_{v,\,i}$ is the intensity from a single particle. There are two types of coherent mechanisms. The first requires the presence of particle bunches of dimensions less than a wavelength, separated by distances greater than a wavelength; when particles are bunched in this way, fields rather than intensities add, so if N is the number of particles per bunch, the resultant intensity is $I_v \approx N^2 I_{v,\,i}$. The second type requires maser amplification, i.e., stimulated emission of radiation. With a maser mechanism the intensity for a source of size l with a coefficient of spontaneous emission j and coefficient of absorption κ is given by

$$I_v = \frac{j}{\kappa}(1 - e^{-\kappa l}). \tag{10-2}$$

If the absorption coefficient κ is negative and $-\kappa l \gg 1$, I_v will increase exponentially with the source size, thus fulfilling the condition in Equation 10-1.

Maser mechanisms account naturally for highly beamed and highly polarized radiation. The coefficient κ is normally a function of direction and polarization, so the intensity I_v is a strong function of these parameters. Coherent emission from particle bunches can also be 100 percent polarized. Consider curvature radiation for example: if the degree of coherence is a maximum in the direction tangent to the field, where the individual particle radiation is completely linearly polarized, then the resultant coherent emission will likewise be 100 percent linearly polarized.

Most models for pulsar emission mechanisms have invoked particle bunching to obtain the observed radiation intensities. Sturrock (1971) pointed out that charge flow along the open field-lines is unlikely to be steady. In Sturrock's pulsar model (p. 184) a cascade of electron-positron pairs forms near the surface of the star. Separation of these charges cancels the accelerating field, so one might expect an oscillatory charge flow with a series of charge sheets leaving the polar cap region of the star. Radiation from these sheets may be coherent at radio wavelengths. An alternative bunching mechanism has been suggested by Goldreich and Keeley (1971). They show that a stream of particles moving on a circular path is unstable to clumping, provided the width of the stream and its velocity dispersion are not too large. Variations in the coherent radiation-reaction force along the stream result in the growth of charge fluctuations with scales l that, when expressed as a fraction of the radius of curvature ρ_c, satisfy the relation $\gamma^{-3} \ll l/\rho_c \ll 1$. For a particle stream with $\gamma \approx 10^4$ and density $\sim 10^{16}$ cm^{-3}, moving along a field line with $\rho_c \approx 10^7$ cm, the growth rate of the clumping instability is large compared to c/ρ_c, and coherent curvature radiation in the radio range could be produced.

In the magnetospheric model of Ruderman and Sutherland (1975) the plasma in the open field-line region consists of secondary electrons and positrons streaming outward with $\gamma_s \approx 10^3$, together with a smaller density of primary positrons streaming outward with $\gamma_p \approx 10^6$ (p. 186). This situation is expected to give rise to the classical "two-stream" instability, in which the Coulomb interaction between the plasma and the fast-moving positrons results in formation of plasma waves of large intensity. In the rest frame of the secondary particles the growth rate of the instability is greatest at the plasma frequency

$$\omega_p' \approx \left(\frac{4\pi n_s' e^2}{m}\right)^{1/2}, \qquad (10\text{-}3)$$

where $n_s' = n_s/\gamma_s$ is the number density of secondary particles in their rest frame. Because of time dilation and Doppler shift, the frequency of these plasma waves in the observer's frame is

$$\omega_p = 2\gamma_s \omega_p', \qquad (10\text{-}4)$$

so from Equations 9-28 and 9-29

$$\omega_p \approx \left(\frac{8e\gamma_p \Omega B_0}{mc}\right)^{1/2} \left(\frac{R}{r}\right)^{3/2}. \qquad (10\text{-}5)$$

For $r = 10^2 R \approx 10^8$ cm, $\gamma_p = 10^6$, $\Omega = 6$ s^{-1}, and $B_0 = 10^{12}$ G, the plasma frequency is about 10^{10} s^{-1}. Ruderman and Sutherland estimate that the growth rate of this instability is sufficient to produce strong bunching within a distance of about 10^8 cm. In this region the magnetic field is almost dipolar, so $\rho_c \approx 10^9$ cm, and the critical frequency for curvature radiation from the secondary particles is $\omega_c \approx 10^{11}$ s^{-1}. Below the critical frequency the intensity of curvature radiation is proportional to $\omega^{1/3}$, whereas above the critical frequency the spectrum drops expo-nentially. Therefore, provided $\omega_p < \omega_c$, the lower modes of curvature radiation are greatly enhanced by bunching. Enhancement factors are estimated to be greater than 10^{17}, leading to brightness temperatures similar to those observed. Spectra rising toward longer wavelengths (as observed) are expected, since coherent radiation is obtained from larger bunches at longer wavelengths. When the bunch size exceeds the wave-length, the coherency factor drops off very rapidly. This effect could result in the observed steepening of pulsar spectra at higher frequencies.

Models involving coherent emission from particle bunches have been generally criticized by Ginzburg and Zheleznyakov (1970) because of the short lifetime of the bunches. For example, with a velocity dispersion $\Delta v = 10^9$ cm s^{-1} and a bunch size $l = 10$ cm, the bunch will dissipate in a time $\Delta t \approx l/\Delta v = 10^{-8}$ s. However, if the bunch is moving toward the observer at highly relativistic speeds, as in the model of Ruderman and Sutherland, then velocity dispersions in the observer frame will not be as large and hence the bunch lifetime will be longer.

In the amplitude-modulated noise model of Rickett (1975) the emitted signal is assumed to consist of shot-noise pulses of duration about $(\Delta v)^{-1}$, where Δv is the total bandwidth of the radiation, i.e., about 10^9 Hz. These nanosecond pulses may be identified with the emission from a single coherent bunch. The time scale could represent either the lifetime of the state of maximum coherence or the time during which a narrow beam is directed toward the observer. Cordes (1975b) has shown that the observed pulse microstructure from PSR 2016+28 may be represented by amplitude modulation of white Gaussian noise. Such noise would (from the central limit theorem) be created by the incoherent addition of many independent shot-noise pulses. This incoherent addition could occur at the source (if many independent bunches contributed to the instantaneous signal), in the pulsar magnetosphere or interstellar medium (because of dispersive and scattering processes), or in the receiver (because of the smoothing resulting from the finite bandwidth).

In contrast to the bunching mechanisms, maser amplification can occur in a uniform medium. To obtain a negative absorption coefficient, an inverted population is required. Relativistic particles with an aniso-tropic velocity distribution constitute such an inverted population. For example, in the pulsar magnetosphere pitch angles (ξ) for particles escaping along the open field-lines are essentially zero, owing to both the short lifetime against synchrotron radiation near the star and conservation of the adiabatic invariant $B/\sin^2\xi$ as the particles move outward. Two basic types of maser amplification may occur: either the electromagnetic waves are directly amplified, or plasma waves are amplified and subsequently converted into electromagnetic waves.

A maser mechanism of the first type, in which the radiation originates at the surface of the neutron star in the vicinity of the magnetic poles, has been proposed by Chiu and Canuto (1971). In an extremely strong magnetic field the energy of an electron (for motion perpendicular to the field direction) is quantized into the Landau levels, so the total energy is given by

$$\varepsilon = \left(k + \frac{1}{2} \right) \hbar \omega_B + p^2/2m, \tag{10-6}$$

where k is a quantum number, $\omega_B = eB/mc$ is the cyclotron frequency, and p is the momentum in the direction of the field. For $B \approx 10^{12}$ G, $\hbar \omega_B \approx 10^4$ eV, so for nonrelativistic electrons only the lowest Landau levels are occupied. The two main radiation processes in such a system are the quantum analog of synchrotron radiation,

$$e(k) \rightarrow e(k') + \gamma, \tag{10-7}$$

and Coulomb bremsstrahlung from collisions with ions,

$$e(k) + Z \rightarrow e(k') + Z + \gamma. \tag{10-8}$$

The latter process can operate when the electrons are in the lowest Landau level ($k = 0$), corresponding to one-dimensional motion along the field, and will generate a continuum of photon energy up to the initial electron energy. Chiu and Canuto have shown that, provided the transition probability does not increase with momentum p faster than the first power and the number of electrons with momentum p increases with increasing p, the absorption coefficient is negative. For the Coulomb bremsstrahlung process the condition on the transition probability is satisfied for nonrelativistic electrons. For a streaming motion with energy of about 10 eV per electron (corresponding to a field of about 0.1 V cm^{-1}), Chiu and Canuto find that the maser saturates in a distance of approximately 1 cm with brightness temperatures in the range 10^{21} to 10^{28} K depending on the detailed parameters.

A second maser mechanism involving amplification of electromagnetic waves has been proposed by Cocke (1973). In this model amplification occurs in regions where electrons are accelerated quasistatically to ultrarelativistic energies by an electric field. For example, electric fields in the polar cap region accelerate electrons along the field-lines to energies as large as 10^{12} eV (p. 184). Cocke finds that for propagation through a region of dimension l containing an electric field E at an angle θ with respect to the electric field, where $\theta\gamma^{1/2} \gtrsim 1$ and

$$\gamma = eEl/(mc^2) \gg 1, \tag{10-9}$$

the optical depth is given by

$$\tau \approx \frac{1}{4\pi}\theta^2\gamma^2(1 - \theta^2\gamma)\left(\frac{\omega_p}{\omega}\right)^2. \tag{10-10}$$

For propagation at $\theta \approx 2\gamma^{-1/2}, \gamma \approx 10^6, \omega_p \approx 10^7\,\mathrm{s}^{-1}$, and $\omega \approx 10^9\,\mathrm{s}^{-1}$, the optical depth is about -100, leading to strong amplification. There is no amplification at angles $\theta \gg \gamma^{-1/2}$ because of the absence of electron motion perpendicular to the field, and thus the emission is strongly beamed in the direction of the magnetic field.

A mechanism for coherent amplification of electromagnetic waves by means of induced scattering of plasma waves has been suggested by Ginzburg, Zheleznyakov, and Zaitsev (1969). These authors consider a cold, i.e., nonrelativistic, plasma in which a large intensity of longitudinal plasma waves at frequencies near ω_p has been excited, e.g., by the two-stream instability. Conversion of these waves to electromagnetic waves at frequencies $\omega \approx \omega_p$ can occur by spontaneous and induced scattering. Ginzburg Zheleznyakov, and Zaitsev find that, for an electron density of about 10^8 cm^{-3}, an energy density of plasma waves greater than 3×10^{-6} erg cm^{-3}, and a path length of about 10^8 cm, the optical depth is large and negative for radiation in the radio frequency range.

In this model the frequency of the scattered radiation is close to the plasma frequency because the plasma is cold. However, the plasma in both the open and closed field-line regions is almost certainly highly relativistic. Radiation intensities at the source implied by the observations provide independent evidence that the plasma in the emitting region is relativistic. Regardless of the emission mechanism, the plasma energy density, U_p, must exceed the energy density of the radiation, U_R; that is,

$$U_p = n\varepsilon > U_R = F/c, \tag{10-11}$$

where F is the radiation flux at the source and n is the density of particles with energy ε. Typically, the radiation flux is greater than 10^{14} erg cm^{-2} s^{-1}, so the plasma energy density must exceed 3×10^3 erg cm^{-3}; hence, for $n \lesssim 5 \times 10^9$ cm^{-3}, the particle (electron) energies must be relativistic. Fluxes are much larger for the Crab pulsar ($F \gtrsim 10^{20}$ erg cm^{-2} s^{-1}), so the density of a nonrelativistic gas would have to be greater than 10^{16} cm^{-3}. Densities this large are unlikely except perhaps close to the star.

A relativistic plasma in which the particles have essentially zero pitch angle is unstable to the growth of plasma waves (Kaplan and Tsytovich, 1973a). In a region of strong magnetic fields ($\omega_B \gg \omega_p$) three different types of wave can exist at low frequencies ($\omega \ll \omega_B$): longitudinal waves with $\omega \approx \omega_p \approx (4\pi ne^2 c^2/\varepsilon)^{1/2}$, Alfvén waves (ordinary mode), and fast magneto-acoustic waves (extraordinary mode). The anisotropic distribution of pitch angles leads to an instability in the Alfvén waves, with the growth rate, Γ_A, given by

$$\frac{\Gamma_A}{\omega_p} \approx \left(\frac{c}{v_A}\right)^{3\gamma} \tag{10-12}$$

where the number of particles with energy ε is proportional to $\varepsilon^{-\gamma}$, and $v_A = cB(4\pi n\varepsilon)^{-1/2}$ is the Alfvén velocity. As $v_A \gg c$ in the region of interest, the growth rate is less than the plasma frequency but it is sufficiently large that the energy density of the Alfvén waves is comparable to the plasma energy density for any source of reasonable dimensions. Maximum growth of Alfvén waves occurs at frequencies close to ω_p, the frequency of the longitudinal waves. Consequently, transfer of energy from Alfvén waves to longitudinal waves by means of Compton or induced scattering is very efficient. Kaplan and Tsytovich (1973b) give for the growth rate of longitudinal waves

$$\frac{\Gamma_L}{\omega_p} \approx \left(\frac{c}{v_A}\right)^4 \frac{U_L}{U_p}, \tag{10-13}$$

where U_L is the energy density of the longitudinal waves. This process will saturate when the energy densities U_A and U_L are comparable to each other and to U_p; the plasma will then be in a highly turbulent state. Several possible mechanisms for conversion of the plasma waves into electromagnetic waves exist. First, the density fluctuations could result in coherent curvature radiation, as suggested by Ruderman and Sutherland (1975). Alternatively, according to Kaplan and Tsytovich (1973b), transfer of energy between longitudinal and transverse waves is very efficient with a growth rate

$$\frac{\Gamma_T}{\omega_p} \approx \frac{U_L}{U_p}. \tag{10-14}$$

Consequently, for $\omega \gtrsim \omega_p$, energy density of electromagnetic radiation is comparable to the plasma energy density and the effective brightness temperature is

$$T_b \approx \frac{2\pi^2 U_R}{k}\left(\frac{c}{\omega}\right)^3 \lesssim \frac{2\pi^2 U_p}{k}\left(\frac{c}{\omega_p}\right)^3. \tag{10-15}$$

If $n \approx 10^{10}$ cm^{-3} and $\varepsilon \approx 10^3$ mc^2, then $\omega_p \approx 2 \times 10^8$ s^{-1} and, from Equation 10-15, the maximum brightness temperature for the emitted radiation is about 5×10^{30} K, adequate to account for the observed radio emission.

Emission from Outside the Light Cylinder

Much of the energy loss from pulsars is expected to be in the form of magnetic-dipole radiation at the rotation frequency (Ω) of the pulsar. Because of the extremely large fields and low frequencies associated with these waves, they are efficient accelerators of charged particles (p. 177). Models in which the pulsed emission is generated outside the light cylinder by these high-energy particles have been proposed by Lerche (1970a,b) and Michel (1971). Because of its low frequency the magnetic-dipole radiation cannot propagate through ionized gas of the density usually found in interstellar space. Therefore the radiation must clear a cavity around the pulsar, the dimensions of which will be determined by pressure balance between the wave and the surrounding medium. Lerche (1970a) suggests that the walls of this cavity oscillate radially at frequency Ω because of the varying pressure of the dipole radiation. Current sheets contained within the wall would then radiate at high frequencies in a way analogous to the synchro-Compton mechanism described by Rees (1971). Lerche finds for the Crab pulsar, at particle densities of 10^2 cm^{-3} and energies of 10^6 mc^2 in the emitting region, a cavity radius $r_c \approx 10^3$ $R_L \approx 10^{11}$ cm and an oscillation amplitude of about 10^4 cm. The emission is coherent at radio frequencies and incoherent in the optical and X-ray regions, and has a continuous spectrum throughout the entire range, with $\alpha \approx -0.67$. Pulse widths are related to the thickness of the interface, and varying orientations of currents throughout the interface are presumed to produce the variations in polarization position angle across pulses.

This model has the advantage that the radiation does not have to escape from the pulsar magnetosphere. However, several problems arise in the model, the most basic of which concerns the stability of the interface. Energy densities and hence pressures are so high within the cavity that there seems to be no means of preventing its expansion. As discussed in Chapter 4 (p. 67), Rees and Gunn (1974) argue that the cavity surrounding the Crab Nebula pulsar currently has a radius of about 5×10^{17} cm, or one-tenth the size of the nebula—much greater than the value obtained by Lerche. Interpulses are difficult to explain with this model, particularly those not occurring midway between the main pulses, because the radiation field would be almost completely dipolar at the interface. Other problems include accounting for the observed long-term stability of integrated pulse profiles and the incorrect prediction concerning the spectrum of the pulsed radiation from the Crab pulsar. As may be seen in Figure 4-2 (p. 59), the spectrum is clearly not continuous.

Michel (1971) suggests that because of interaction of the intense wave fields with the plasma the waves close to the light cylinder become "square" rather than sinusoidal, and form regions of opposite field polarity separated by neutral sheets. Plasma then collects at these neutral sheets, forming a shock wave in the process. Coherent radio emission could result from the collective motion of these particles; for the Crab pulsar normal synchrotron emission from the same particles could produce the observed optical radiation. As the sheets expand with a velocity near c, an observer would see significant radiation only from those regions moving toward him. Therefore, since the sheets are separated by a distance c/Ω, the radiation would appear to be pulsating at the frequency Ω. Again, it is difficult to account for the stability and polarization of the observed pulse profiles by this mechanism.

Emission from the Closed Field-Line Region

Most models of the pulse emission mechanism consider a source within the co-rotating zone of the magnetosphere that, at least to an outside observer, produces a radiation beam of fixed orientation with respect to the star. In one of the earliest such models, Eastlund (1968, 1970) proposed that the integrated pulse profile be identified with the radiation pattern of synchrotron emission from particles bunched in the equatorial region of the pulsar magnetosphere. For certain particle energies and pitch angles, low-order modes of synchrotron radiation have a double-lobed beam pattern, and would produce a two-component pulse as the star rotates. The separation of these components would vary with frequency in a manner similar to that observed. The observed stability of integrated profiles requires that particle pitch angles and energies remain constant over long periods of time; however, it seems unlikely that the particle parameters would have such stability because the plasma in the pulsar magnetosphere is almost certainly highly turbulent. Furthermore, as discussed earlier in this chapter (p. 202), observations suggest that the entire integrated profile is not generated by a single group of particles.

To an observer located far from the star, the radiation from an emission region located just inside the light cylinder would be compressed into a narrow pulse and enhanced in intensity by relativistic effects. As for synchrotron radiation, a source that is quasi-isotropic in the co-rotating frame would have an effective beamwidth of order γ_ϕ^{-3} in the longitude direction and γ_ϕ^{-1} in the latitude direction, so that the source intensity would be enhanced by a factor γ_ϕ^4. Because of these attractive features many proposed emission mechanisms are based on an emission region located close to the light cylinder. If the radiation beaming is primarily due to the co-rotation velocity, the beam is directed parallel to the equatorial plane and tangential to the velocity. Different integrated profiles can be accounted for by different distributions of emitting plasma around the circumference of the magnetosphere. It can be seen from symmetry arguments that such a distribution is not compressed by relativistic co-rotation.

Observations show that the observed widths of subpulses are not strongly dependent on frequency. This fact led Smith (1970, 1973) to propose that subpulses were formed by relativistic beaming with γ_ϕ in the range 2–5. For an isotropic source this would result in subpulse widths of 0.5° to 10° longitude, about the observed values. The width of the beam in latitude would be 10° to 30°, and thus, for a random orientation of rotation axes, between 10 and 30 percent of all pulsars could (in principle) be observed.

The relations describing this beaming process have been derived by Zheleznyakov (1971) as follows. Consider a frame A' in which a quasi-isotropic source of intensity I'_ω is fixed, and a laboratory frame A in which the source is moving with velocity βc. If θ is the angle between the propagation direction and the velocity in A, then

$$\omega = \frac{(1 - \beta^2)^{1/2}}{1 - \beta \cos \theta} \omega' \tag{10-16}$$

and

$$I_\omega = \frac{(1 - \beta^2)^{3/2}}{(1 - \beta \cos \theta)^2} I'_\omega. \tag{10-17}$$

For the power-law spectrum in the source frame,

$$I'_\omega = K(\omega')^\alpha, \tag{10-18}$$

the intensity in the laboratory frame is

$$I_\omega = \frac{(1 - \beta^2)^{(3 - \alpha)/2}}{(1 - \beta \cos \theta)^{(2 - \alpha)}} K\omega^\alpha. \tag{10-19}$$

Therefore the spectrum of a source with a power-law spectrum in its rest frame remains power-law (with the same spectral index) in the laboratory frame. The beamwidth ($\Delta\theta$) of the radiation in the laboratory frame is determined by the factor $(1 - \beta \cos \theta)^{2 - \alpha}$; hence, for a source with a power-law spectrum the beamwidth is independent of frequency. In the laboratory frame, from Equation 10-19 and assuming $1 - \beta \ll 1$,

$$\Delta\theta \approx 2\eta(1 - \beta^2)^{1/2} \tag{10-20}$$

where $\eta^2 = 2^{1/(2-\alpha)} - 1$. As is true for synchrotron radiation, the pulse width in longitude as seen by an outside observer ($\Delta\phi$) is reduced by a factor $(1 - \beta)$ as the source approaches in its circular path, so that in the observer's frame

$$\Delta\phi \approx \eta(1 - \beta^2)^{3/2}. \tag{10-21}$$

This relation is consistent with the approximate beamwidth γ_ϕ^{-3} given above. Likewise, the beamwidth in the latitude direction is, from Equation 10-20, approximately γ_ϕ^{-1}.

Gold (1969) has proposed a similar model in which the emission comes from a source much closer to the light cylinder, with γ_ϕ of 10^5 to 10^6. For Lorentz factors this large the intensity enhancement is very significant. The radiation is assumed to originate in bunches of particles that bounce between the magnetic poles at relativistic speeds; to an outside observer, significant radiation occurs only at the point where the field-lines make their closest approach to the light cylinder. Because of the relativistic motion along the field-lines, the effective latitude width of the radiated beam is much greater than γ_ϕ^{-1}, so a reasonable fraction of pulsars can be seen.

Several different emission mechanisms have been proposed for the radiation sources in these models. Since the beaming is due to relativistic motion, the emission can be quasi-isotropic in the source frame; however, it must have the correct spectrum and produce radiation of sufficient intensity. Because of the relativistic compression of angles, the directional dependence of polarization is compressed in the observer frame. The edges of the observed pulse correspond approximately to $\theta' = \pm\pi/2$. Consequently, if the position angle of radiation in the source frame is perpendicular to the radius vector and at an angle δ to the velocity vector, then the observed position angle rotates by $\pi - 2\delta$ through the pulse. In this way, position angle variations similar to those observed can be obtained. Smith (1970) has suggested that the radiation mechanism in the source frame is cyclotron emission from bunches of electrons, whereas Ginzburg, Zheleznyakov, and Zaitsev (1969) proposed a maser mechanism.

Emission mechanisms in which the source is located near the light cylinder have the advantage of a natural beaming process. There are some disadvantages, however. First, the magnetic field in the vicinity of the light cylinder is a strong function of pulsar period; for a dipole field, B_L is proportional to P^{-3}, hence the magnetic energy density is proportional to P^{-6}. Since pulsar periods cover a range of over two orders of magnitude, one would expect the pulse characteristics of slow pulsars to be quite different from those of fast pulsars, whereas in fact the basic properties of pulsars are remarkably independent of period. For example, as mentioned earlier, and shown in Figure 10-1, there is only a very weak dependence of radio luminosity on pulsar period. Also, observed widths of integrated profiles are, on the average, proportional to period (Figure 2-3, p. 18), implying that the radiated beam has a width independent of period.

The fact that integrated profiles are narrow compared to the period requires that the emission region be confined to a similarly narrow region of the circumference of the light cylinder. However, because of the radiation-reaction force, radiating particles will remain in the emission region for only a short time. Hence a mechanism for continuously supplying energetic particles to a confined region far from the star must be found.

As mentioned above, one of the original motivations for adopting the relativistic beaming model was the frequency independence of the beamwidth. This frequency independence is illustrated in Figure 10-3, in which

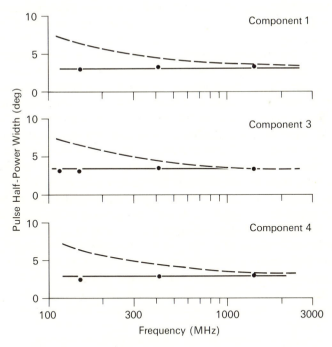

10-3 Mean observed widths of individual pulses from PSR 0329 + 54. The dashed lines indicate widths predicted from the relativistic beaming model assuming isotropic emission in the source frame. The observed points are better represented by a constant pulse width, indicated by the solid lines. [From Manchester, Tademaru, *et al.,* 1973.]

observed widths of subpulses from PSR 0329 + 54 at several frequencies are plotted. However, because of the factor η in Equation 10-21, the predicted pulse widths are dependent on the spectral index of the radiation. with more positive indexes giving larger widths. (This effect arises because in the source frame the wings of the pulse are emitted at a higher frequency than the central part of the pulse.) The spectrum of PSR 0329 + 54 has a low-frequency cutoff, with peak intensity at about 400 MHz (Figure 2-6, p. 23); provided this cutoff is intrinsic, the pulse widths expected for relativistic beaming are not constant. Widths predicted from the observed spectral index are indicated by dashed lines in Figure 10-3. Ginzburg and Zheleznyakov (1975) have pointed out that the lack of agreement between the observed and predicted widths could be avoided if the emission in the source frame was nonisotropic and/or the source location (and hence β) varied in such a way as to compensate for the varying relativistic effect.

One of the outstanding characteristics of pulsar emission is the great stability in both shape and relative phase of the pulsars' integrated profiles, many of which are rather complex, and their polarization. This observed stability suggests that particle motions are determined by the magnetic field, that the fields in turn are frozen into the solid crust of the neutron star, and (except for models such as those described on p. 211f.) that the emission source is located within the co-rotating zone of the pulsar magnetosphere. The co-rotating zone must stop short of the light cylinder. If the Alfvén velocity is less than c at the light cylinder, co-rotation breaks down at the *Alfvén cylinder*, i.e., the radius where the co-rotation velocity equals the Alfvén velocity. As described in Chapter 9 (p. 187), Henriksen and Rayburn (1974) find that this condition limits the co-rotation Lorentz factor (γ_ϕ) to less than $\sqrt{2}$, or β to less than 0.71. Other authors suggest that co-rotation ceases even closer to the star; for example, Hinata and Jackson (1974) find that co-rotation breaks down at $\beta \approx 6 \times 10^{-3}P^{-3/5}$, where P is the period in seconds. For the Crab pulsar this corresponds to only five percent of the light-cylinder radius, and for longer-period pulsars even less. If co-rotation is restricted to such a limited region, it is extremely unlikely that models in which the emission region is located near the light cylinder can account for the observed stability of pulse shapes and phases.

The Alfvén-cylinder condition is equivalent to requiring that the (formal) Alfvén velocity exceed the velocity of light in the emission region, or that the energy density of the magnetic field, $U_B = B^2/8\pi$, exceed the energy density of the emitting plasma, U_p. Also, U_p must exceed U_R, the radiation energy density (Eqn. 10-11), so we have the condition that in the emitting region.

$$U_B \gg U_R. \tag{10-22}$$

From Figure 10-1 we see that U_R (computed assuming an effective source area of 10^{15} cm^2) exceeds U_B in the vicinity of the light cylinder by a large factor for many of the longer-period, highly luminous pulsars. This source area corresponds to a light travel time of 1 ms; in fact, many pulsars have microstructure with time scales at least an order of magnitude smaller than this, making the radiation energy densities at least two orders of magnitude larger than those given in Figure 10-1. However, there may be some reduction in the radiation energy densities if relativistic beaming is important.

For the Crab and other short-period pulsars, the condition in Equation 10-22 is satisfied everywhere within the light cylinder. For the longer-period pulsars, however, the energy density arguments together with the limitations on the size of the co-rotating zone appear to rule out location of the emission source near the light cylinder. For example, if we take $\beta = 0.9$ for PSR 1133 + 16, then from Equation 9-6 (p. 177) $B \approx 16$ G and $U_B \approx 10$ erg cm^{-3} in the emitting region. Some micropulses in this pulsar have widths of less than 100 μs and flux densities of 10^3 Jy or 10^{-20} erg cm^{-2} s^{-1} Hz^{-1}. If we assume that the microstructure has a bandwidth of 100 MHz, the implied radiation flux at the source is greater than 10^{16} erg cm^{-2} s^{-1}. Because of the enhancement resulting from relativistic beaming, this flux must be reduced by a factor $(1 - \beta^2)^2/(1 - \beta)^4 \approx \gamma_\phi^4 \approx 30$ for $\beta = 0.9$ (Manchester, Tademaru *et al.*, 1973). Therefore, in the laboratory frame $U_R = F/c \approx 10^4$ erg cm^{-3}. Clearly, the condition in Equation 10-22 is not satisfied; the conclusion is that if the emission source is located at $\beta = 0.9$, the characteristics of the profile cannot be determined by the magnetic field. Manchester and co-workers have shown that, with typical pulsar parameters, U_R exceeds U_B throughout the range $0.2 < \beta < 0.98$. At small values of β, close to the star, the magnetic field rapidly increases in strength; at high β, close to the light cylinder, the beaming enhancement factor becomes large. The high-β region, however, may be ruled out as a location for the emission source by the breakdown of co-rotation.

Emission from the Open Field-Line Region

The narrow integrated profile observed for most pulsars implies that the emission originates from a confined zone on the rotating neutron star. Observations of the Vela pulsar (PSR 0833 − 45) by Radhakrishnan and co-workers (1969) showed that the position angle of the linearly polarized emission changes by more than 50° across the integrated profile. Furthermore, no differential Faraday rotation was observed, showing that, at a given point in the profile, all frequencies are emitted at the same position angle—an angle presumably related to the magnetic field direction in the source region. The observed rapid change in position angle implies therefore that the source region may be in the vicinity of a magnetic pole. This conclusion was further supported by the work of Goldreich and Julian (1969), who showed that charged particles would be accelerated along the open field-lines emanating from the polar regions. As originally proposed by Radhakrishnan and Cooke (1969), these particles would emit radio-frequency curvature radiation (p. 184) in the direction of their motion, the high intensities being obtained by coherence effects within particle bunches. This radiation would be expected to form a conical beam, directed radially outward from the star and centered on the magnetic axis; the angular extent of the cone is determined by the angle subtended by the open field-lines in the source region (Figure 10-4). For emission near the surface and an approximately dipolar field, the cone half-angle (θ_b) is $\sin \theta_b \approx \sin \theta_p = (\Omega R/c)^{1/2}$ (see Eqn. 9-15, p. 180). The emission cone is hollow because, at least to a first approximation, field-lines close to the magnetic axis are straight, so there is no curvature radiation in this direction. Since curvature radiation is beamed in the direction tangential to the field, different parts of the pulse are emitted from different parts of the polar cap. If the locus of the emitting region (the line PS in Figure 10-4) passes close to the magnetic axis, the pulse profile will be double-peaked in form, as is often observed. The observed pulse width in longitude, $\Delta\phi = 2\phi_p$, where ϕ_p is given by

$$\cos \phi_p = \frac{\cos \theta_b - \cos \alpha \cos \zeta}{\sin \alpha \sin \zeta} \qquad (10\text{-}23)$$

and α and ζ are the angles made by the rotation axis with the magnetic axis and observer direction, respectively.

Curvature radiation is polarized parallel to the plane of curvature of the field, which, for an undistorted dipole field, contains the magnetic axis. Therefore, radiation from the point P in Figure 10-4 has a position angle ψ relative to the projected direction of the rotation axis (the meridian RP), i.e., it will be polarized parallel to the projected direction of the magnetic axis. From the spherical triangle PQR, one obtains

$$\tan \psi = \frac{\sin \alpha \sin \phi}{\sin \zeta \cos \alpha - \cos \zeta \sin \alpha \cos \phi}. \qquad (10\text{-}24)$$

The rate of change of position angle reaches a maximum value

$$\left(\frac{d\psi}{d\phi}\right)_{max} = \frac{\sin \alpha}{\sin(\zeta - \alpha)} \qquad (10\text{-}25)$$

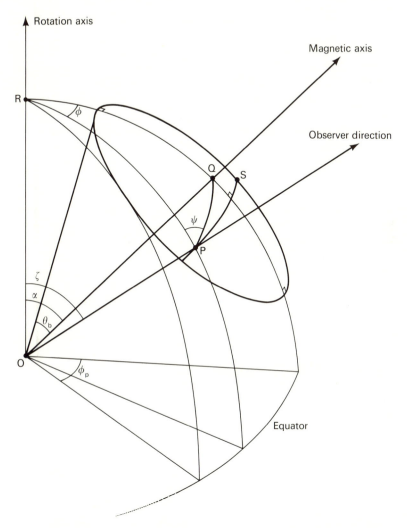

10-4 Axes and angles used to describe the magnetic-pole model for pulse emission.
At any given instant the observer sees emission from the point where the
field-lines are tangential to the line of sight (point P in the figure). As the star
rotates, P moves along the arc PS.

when the line of sight crosses the meridian containing the magnetic axis, where $\phi = 0$. Clearly $(d\psi/d\phi)_{max}$ is greatest when ζ is close to α, i.e., when the observer's line of sight passes close to the magnetic axis. Observed position-angle curves for four pulsars, together with the least-squares best fit of Equation 10-24 to each set of data, are shown in Figure 10-5. The shape of the fitted curve is not sensitive to the value of α provided it is not small; $\alpha = 60°$ was assumed. It is clear that the rotating-vector model gives an extremely good fit to the position angle data. Similar variations in position angle are observed across the optical main pulse and interpulse of the Crab pulsar (see Figure 4-8, p. 74). If the pulse emission is polarized parallel to the projected field direction, then the (intrinsic) position angle at the center of symmetry of the position-angle curve is that of the stellar rotation axis projected on the sky.

This model has been further developed by a number of authors. Komesaroff (1970) discussed the frequency dependence of the integrated pulse profile and the emission intensity on the assumption that the radio pulses were coherent curvature radiation from thin charged-particle

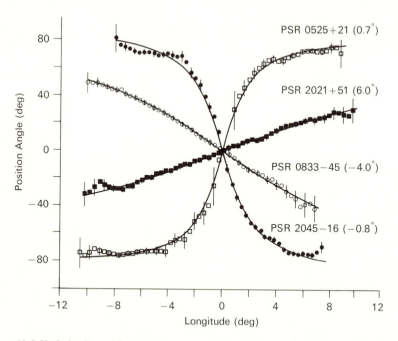

10-5 Variation in position angle throughout the integrated profiles of four pulsars together with best-fit curves based on the rotating-vector model (Equation 10-24). The minimum angular separation of the observer's line of sight from the magnetic axis $(\zeta - \alpha)$ is given for each curve.

sheets leaving the polar cap region. Variation with frequency in the number of particles radiating in phase, i.e., the size of the first Fresnel zone, results in a frequency dependence of $\omega^{-1/4}$ in the separation of the components of double profiles, a dependence similar to that observed (see Figure 2-4, p. 19). The predicted spectral index of the coherent emission is -1, in reasonable agreement with the observations. A similar analysis by Tademaru (1971), based on Sturrock's (1971) model for the polar regions (p. 184), gives the same frequency dependence for the separation of double-pulse components. For a thin sheet of radiating charges (thickness $d < \lambda$), Tademaru found that the spectral index varies from -0.5 at low frequencies to -2 at high frequencies, whereas for a thick sheet ($d > \lambda$) the spectral index varies between -1.5 and -4 from low to high frequencies. These values encompass the observed range of spectral indices (see Figure 2-7, p. 23).

For a strictly dipolar field, the angular width of the cone of open field-lines near the stellar surface is smaller than the observed pulse widths for most pulsars. However, the last closed field-line almost certainly does not extend all the way to the light cylinder, so the open field-lines subtend an angle larger than $\theta_p \approx (\Omega R/c)^{1/2}$ at the star. Alternatively, within the constraints discussed in the previous section, the emission region may be located farther from the star.

Ruderman and Sutherland (1975) have described a fairly complete model for the pulse emission process based on their theory of magnetospheric gaps (p. 185). They propose that the secondary electron–positron pairs radiate curvature radiation at radio frequencies at distances (for a pulsar of about one-second period) of about 10^8 to 10^9 cm from the star. At these distances the field is nearly dipolar and the angle subtended by the open field-lines is close to the observed pulse widths. For field-lines that originate near the polar cap, the radius of curvature is given by

$$\rho_c = 4r^2/3a, \tag{10-26}$$

where r is the distance from the neutron star and a is the perpendicular distance from the magnetic axis. From Equation 9-23 (p. 184) the frequency of curvature radiation is then

$$\omega_c = \frac{9\gamma_s^3 ac}{8r^2}. \tag{10-27}$$

Such radiation can only occur on the open field-lines. From Equation 9-15 (p. 180) the outer edge of the emission region is therefore defined by

$$a \lesssim r(\Omega r/c)^{1/2}. \tag{10-28}$$

As discussed earlier (p. 207), significant coherent radiation can only occur when ω_c exceeds ω_p, the plasma frequency. Therefore, from Equations 10-5 and 10-27 the inner edge of the emission region is defined by

$$a \gtrsim \frac{16}{9\gamma_s^3 c}\left(\frac{2e\gamma_p\Omega B_0 R^3}{mc}\right)^{1/2} r^{1/2}. \tag{10-29}$$

These limits on the extent of the emission region are illustrated in Figure 10-6. In this model higher frequencies are generated nearer the star and lower frequencies farther out. Ruderman and Sutherland show that

$$\omega_{max} \approx 10^{10}P^{-2.4} \quad (\text{s}^{-1}) \tag{10-30}$$

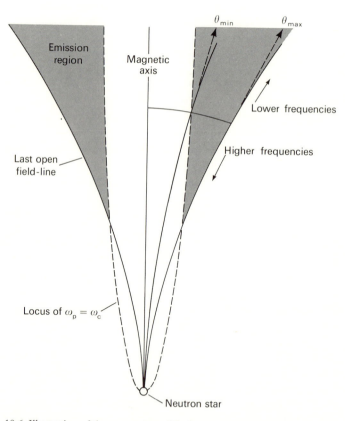

10-6 Illustration of the parameters of Ruderman and Sutherland's pulsar model (1975).

and that the corresponding $r_{min} \approx 10^8$ cm for typical pulsar parameters. Particle bunching by the two-stream instability would be expected out to distances of at least 10^9 cm, so from Equation 10-5 a low-frequency cutoff would occur at $\omega_{min} \approx 10^8$ s^{-1}. These values are in reasonable accord with observations, although there does not seem to be as strong a period dependence of the upper-frequency spectral cutoff as that predicted by Equation 10-30.

A dipole field line makes an angle

$$\theta = 3a/2r \tag{10-31}$$

with the magnetic axis. Consequently, for radiation at a given frequency, Equations 10-5, 10-28, and 10-29 show that

$$\theta_{max} = \frac{3}{2}\left(\frac{2\Omega}{c}\right)^{1/2}\left(\frac{e\gamma_p\Omega B_0 R^3}{mc}\right)^{1/6}\omega^{-1/3} \tag{10-32}$$

and

$$\theta_{min} = \frac{8}{3\gamma_s^3 c}\left(\frac{e\gamma_p\Omega B_0 R^3}{mc}\right)^{1/3}\omega^{1/3}. \tag{10-33}$$

Substituting reasonable values for the various quantities in these equations, Ruderman and Sutherland find that

$$\theta_{max} \approx 16P^{-0.7}\omega_{10}^{-1/3} \quad \text{(degrees)} \tag{10-34}$$

and

$$\theta_{min} \approx 16P^{0.9}\omega_{10}^{1/3} \quad \text{(degrees)} \tag{10-35}$$

where ω_{10} is the angular radio frequency in units of 10^{10} s^{-1}. Current densities are likely to be greatest near the outer edge of the open field-line region, and therefore Equation 10-34 predicts that the separation between components of double profiles should vary approximately as $\omega^{-1/3}$, in reasonable agreement with observations. Equation 10-35 shows that these components should be more clearly separated in long-period pulsars. This also is in accord with observations; nearly all type C pulsars have periods in excess of one second. These equations also imply that components should be narrower at higher frequencies; there is some, albeit weak, observational evidence for this.

When an electron–positron cascade commences at some point in the polar gap, the local value of $\mathbf{E} \cdot \mathbf{B}$ is reduced, thereby inhibiting discharge in adjacent regions. Consequently, one would expect the discharge to be confined to a few localized regions or "sparks." These sparks would inject a stream of positrons into the magnetosphere, giving rise to radio emission. The lifetime of sparks is expected to be about 10 μs, so the associated bursts of emission could be identified with microstructure. A group of such bursts would constitute a subpulse. Since different frequencies are emitted at different distances from the star, the observed longitude of a subpulse would vary with frequency, being closer to the pulse center at high frequencies. This is in accordance with observation.

If the emission region is much closer to the star than to the light cylinder, the effects of propagation through the pulsar magnetosphere into the interstellar medium must be considered. In such models the cyclotron frequency in the emission region is much greater than the observed radio frequencies, and in some models even the local plasma frequency may exceed the observed frequencies. Consequently, one would expect propagation effects to be important.

In the model of Chiu and Canuto (p. 208) the emission is generated at the stellar surface by means of an electron-bremsstrahlung process. Virtamo and Jauho (1973) have developed this model further by considering amplification of the ordinary magneto-hydrodynamic mode for which the refractive index in the limit $\omega_B \to \infty$ is given by

$$\mu^2 = \frac{1 - X}{1 - X \cos^2 \theta}, \tag{10-36}$$

where $X = \omega_p^2/\omega^2$. Since this mode is evanescent where $X > 1$, the observed emission must be generated in regions where $\omega_p < \omega$. For the electron-bremsstrahlung process, emission in this mode is greatest in the direction $\theta = \pi/2$. However, because of the variation in refractive index throughout the magnetosphere, all emission generated in a layer characterized by the parameter X is, on exit from the magnetosphere, confined to a cone of half-angle θ_b about the magnetic axis, where $\cos^2 \theta_b = X$. This cone is very narrow for generation in regions where $X \approx 1$. Virtamo and Jauho find that the angle θ_b varies with frequency approximately as $\omega^{-0.45}$, a somewhat stronger variation than the observed frequency dependence of pulse width.

Propagation effects resulting from the shearing motion that must occur near the light cylinder have been considered by Lerche (1974, and in other papers in this series). For a refractive index significantly different from unity, electromagnetic waves would be deflected by the shearing motion and in some cases could not penetrate the medium. These effects would destroy the beaming of any emission source within the light cylinder. However, as pointed out by Elitzur (1974), the refractive index is very close to unity for a relativistic plasma. For example, for longitudinal propagation through a plasma moving along field-lines with Lorentz factor γ, the refractive indexes for the two electromagnetic modes are given by

$$\mu^2 = 1 \pm \frac{1}{2\gamma^2} \frac{X}{Y}, \tag{10-37}$$

where $Y = \omega_B/\omega$. If $\gamma \approx 10^3$ (as in the Ruderman–Sutherland model), then μ is very close to unity, and refractive effects in the vicinity of the light cylinder are unlikely to be important. Furthermore, recent magnetospheric models (p. 187) suggest that co-rotation ceases some distance inside the light cylinder. If this is the case, shear motions would be much less than those assumed by Lerche.

Another factor that must be considered in any model that locates the emission region relatively close to the star is the effect on polarization of the wave as it propagates through the magnetosphere. Close to the star $\omega_B \gg \omega$, so the propagation would be "quasi-transverse" with each of the two modes linearly polarized. However, in the interstellar medium $\omega_B \ll \omega$, so the propagation would be "quasi-longitudinal" with circularly polarized modes. The fact that in most pulsars the observed polarization is predominantly linear suggests that a single mode is generated in the quasi-transverse region. If the variation in propagation conditions from quasi-transverse to quasi-longitudinal is gradual along the ray path, then the emergent polarization would be circular. Because extensive circular polarization is not observed, there must be a region of "limiting polarization" located within the quasi-transverse region. (For a discussion of limiting polarization, see Budden, 1961.)

Zheleznyakov (1970) has suggested that the region of limiting polarization is located well above the emission region in the pulsar magnetosphere. However, as Komesaroff, Ables, and Hamilton (1971) point out, this assumption leads to several problems. Observations over a wide range of frequencies (more than three octaves in some cases) have shown that the variation in position angle across the integrated profile is independent of frequency. This implies that the region of limiting polarization is the same for all frequencies, which seems unlikely if the region is located far from the star. Since observed position angle variations are symmetrically located within the pulse profile, particularly in Type C pulsars, there must be a very close relationship between the observed polarization and the emission region. Furthermore, no phase shift of the polarization relative to the integrated pulse profile, which might be expected if the polarization were imposed at large distances from the star, has ever been observed. These observations suggest that the limiting polarization is determined either within or immediately above the emission region, with the transition occurring at essentially the same place for all frequencies. This would be achieved if the density of free charges were very low throughout the magnetosphere or if relativistic streaming reduced the effect of the charges (see Eqn. 10-37).

Cocke and Pacholczyk (1976) have suggested that mode coupling could produce the orthogonally polarized subpulses described in Chapter 3 (p. 51), as well as the circular polarization observed in individual pulses. For a linearly polarized wave incident (at an arbitrary position angle) on a thermal plasma of density $n_e \approx 10 \text{ cm}^{-3}$ and thickness about 5×10^9 cm, Cocke and Pacholczyk find that the position angle on exit is preferentially aligned either parallel or perpendicular to the projected magnetic field. Especially for incident angles close to 45° and 135°, the fractional linear polarization is reduced and substantial circular polarization created.

An alternative explanation for the orthogonal pulses, again involving propagation effects, has been suggested by Blandford and Scharlemann (1976). They consider a source that emits radiation partially polarized in the direction parallel to the projected magnetic field. Longitudinal scattering of radio photons by streaming electrons affects only photons polarized parallel to the field; thus, if such scattering is strong, only photons polarized perpendicular to the field will penetrate the scattering region. On the other hand, if the scattering is weak, the intrinsic polarization of the source will be seen.

At some point within the magnetosphere of most pulsars the radio frequency (in the rest frame of the electrons) will be equal to the electron cyclotron frequency and cyclotron absorption will occur. Since the observed pulses do not in general have strong circular polarization, if cyclotron absorption is important, approximately equal numbers of electrons and positrons must contribute. Blandford and Scharlemann (1976) find that, with the parameters of the Ruderman–Sutherland model, optical depths are high unless the streaming particles are confined to a small fraction of the open field-lines.

Optical, X-Ray, and γ-Ray Pulse Emission

Until recently, pulsed emission at frequencies above the radio range has been observed only from the Crab pulsar; therefore, most models for the generation of this high-frequency emission have been discussed with reference to this source. The observed differences in the pulse shapes at radio and optical frequencies and the apparent spectral discontinuity between these two regions (Figure 4-2, p. 59) suggest that the radio and optical emission processes are different. Further support for this idea is provided by the lack of observed intensity fluctuations and the much lower brightness temperature of the optical and X-ray emissions. However, the coincidence in time of the radio, optical, and X-ray peaks shows that the emission processes in these different regimes are closely related.

The broadband, nonthermal character of the optical and X-ray spectra, together with the observation of linear polarization of the optical pulses, has suggested a synchrotron emission process to many authors. Because of the similarity between the spectral indexes of the nebular radio emission and the pulsar X-ray emission, Shklovskii (1970) suggested that the same electrons were responsible for the two types of emission. This implies that the transverse magnetic field in the emitting region is about 5×10^3 G, which is consistent with emission near the light cylinder and electron pitch angles of 10^{-2} radians. If the spectral turnover at optical and infrared frequencies is due to synchrotron self-absorption, the required density of relativistic electrons is $\sim 10^{14}$ cm^{-3}, giving an energy density of $\sim 10^{11}$ erg cm^{-3} in an emission region of dimensions $\sim 10^7$ cm. Because of the small pitch angles, the emission is directed outward from the star, almost tangentially to the open field-lines; therefore, provided the pulsar radio emission was similarly beamed along the open field-lines, the radio and optical pulses would be coincident.

Because of the short lifetime against synchrotron radiation (10^{-6}–10^{-4} s) and because of conservation of the transverse adiabatic invariant, electrons accelerated near the star have essentially zero pitch angle in the vicinity of the light cylinder. Therefore, they must acquire their transverse momentum within the emission region. Hardee and Rose (1974) have proposed that, in fast pulsars, transverse plasma instabilities will occur near the light cylinder. The resulting field deformations could provide the transverse field components necessary for synchrotron radiation.

The time-averaged luminosity of the optical pulses from the Vela pulsar is approximately 10^5 times smaller than the corresponding quantity for the Crab pulsar (Wallace *et al.*, 1977). If this luminosity difference is entirely due to the longer period of the Vela pulsar, then optical luminosity is approximately proportional to P^{-12}. Pacini (1971) has shown that for incoherent synchrotron emission from a region near the light cylinder one would expect a P^{-10} dependence of the optical luminosity, because of decreasing magnetic field strengths and particle energies as the light cylinder expands. Within the uncertainties, this model is in reasonable agreement with the observations. A period dependence as strong as P^{-12} implies that the optical luminosity of the Crab pulsar should be decreasing by about 0.5 percent per year. Wampler (1972) has examined photographs taken over an 80-year period and concluded that any change in brightness was less than a factor of two, i.e., less than 0.9 percent per year.

The very sharp peak of the Crab optical pulses (see p. 72) implies that at least some of the particles emit a beam of width less than about 10^{-2} radians. If the optical emission is synchrotron radiation, this narrow beamwidth implies particle Lorentz factors $\gamma \gtrsim 10^2$ and pitch angles $\xi \lesssim 10^{-2}$ radian. O'Dell and Satori (1970) and Epstein and Petrosian (1973) have pointed out that the normal synchrotron relations do not apply for small pitch angles, that is, $\xi \lesssim \gamma^{-1}$. For such small-angle emission the observed intensity peaks at a frequency of

$$\omega_c = 2\omega_B\gamma. \tag{10-38}$$

Therefore, to obtain a peak at optical frequencies ($\omega_c \approx 10^{15}\ \text{s}^{-1}$) with $\gamma \gtrsim 10^2$, we require $\omega_B \lesssim 5 \times 10^{12}\ \text{s}^{-1}$ or $B \lesssim 3 \times 10^5$ G. The field strength everywhere within the light cylinder of the Crab pulsar is greater than this upper limit. Another problem for the synchrotron model concerns circular polarization. For small pitch angles the radiation should be almost totally circularly polarized, whereas observations show less than 0.03 percent intrinsic circular polarization. This objection would not apply, of course, if the radiation were generated by an equal mixture of positrons and electrons.

A related model proposed by Zheleznyakov and Shaposhnikov (1972) and Smith (1973) overcomes the magnetic field strength problem. In this model emission is by the synchrotron process, but the beaming results from relativistic co-rotation velocities. In this case the apparent beamwidth in longitude is proportional to γ_ϕ^{-3} (p. 214) so the intrinsic beamwidth of the emission process may be much larger. Zheleznyakov and Shaposhnikov find that the observed optical luminosity and spectrum of the Crab pulsar could be produced by a source of dimensions about 5×10^7 cm located about 10^8 cm from the star. (The radius of the light cylinder is about 1.6×10^8 cm.) Portions of the source close to the light cylinder, where $\gamma_\phi \approx 5 - 10$, could produce the observed cusp-like peak to the pulse. Required electron densities are about $5 \times 10^{11}\ \text{cm}^{-3}$, with an energy spectrum having an index of about -1.4 below 3×10^8 eV and -3.4 above this energy. A limit can be placed on the energy density (U_p) of the relativistic electrons based on the lack of an excess of high-energy γ-ray radiation. Such an excess would be produced by inverse Compton scattering of the optical and X-ray photons if the ratio $U_B/U_p \lesssim 30$ in the emission region. This limit also ensures that the motion of the electrons is controlled by the magnetic field in the emission region. In this model, too, strong circular polarization is expected, unless the emission is generated by an equal mixture of electrons and positrons or the magnetic field in the emission region is within a few degrees of orthogonal to the line of sight (Epstein and Petrosian, 1973). The model assumes co-rotation of the plasma in the vicinity of the light cylinder and would not be compatible with radial beaming of the radio emission.

The above discussion suggests that the optical pulses from the Crab pulsar may not be generated by the synchrotron process. Alternate emission processes have been proposed by Kaplan and Tsytovich (1973a) and Sturrock, Petrosian, and Turk (1975). As described earlier (p. 210), a relativistic plasma in which the particles have essentially zero pitch angle is unstable to the growth of plasma waves. Kaplan and Tsytovich suggest that the optical emission results from inverse Compton scattering of plasma waves by the relativistic particles. The frequency of the scattered radiation is given by

$$\omega \approx 2\omega_p\gamma^2, \tag{10-39}$$

where γ is the Lorentz factor for the relativistic particles. For $\omega_p \approx 10^9\,\text{s}^{-1}$ and $\gamma \approx 10^3$ (as in the Ruderman–Sutherland model), the frequency of scattered radiation would be approximately $10^{15}\,\text{s}^{-1}$, that is, in the optical range.

In the pulsar model proposed by Sturrock (1971) and developed by Ruderman and Sutherland (1975), primary particles of energy 10^6 to $10^7 mc^2$ radiate γ-ray photons capable of pair-production in the lower magnetosphere (p. 184). About 10^3–10^4 secondary particles of energy 10^3–$10^4 mc^2$ are produced for every primary particle. Sturrock, Petrosian, and Turk (1975) propose that the secondary particles generated by a given primary particle form a coherent bunch immediately after their creation, radiating curvature radiation at optical frequencies. For significant coherence the dimensions of the bunch in the direction of propagation must remain less than the wavelength of the optical radiation, and in the perpendicular direction less than $\gamma_s\lambda$ for a time of order 10^{-5} s. Estimates show that these requirements are not unreasonable. Each bunch initially contains an equal number of electrons and positrons; for full coherence these two species must separate by more than a wavelength. An electric field in the emission region of 15 V cm^{-1} would be sufficient to achieve this separation.

The observed lack of intensity fluctuations is often used as an argument against coherent emission processes at optical frequencies. However, in this coherent emission model the radiation from different bunches adds incoherently, and since the number of bunches radiating at a given time is estimated to be about 10^{29}, statistical fluctuations would be negligible. The optical luminosity obtained by Sturrock, Petrosian, and Turk is proportional to $P^{-9/2}$. This is a much weaker dependence than the P^{-12} dependence observed. As the optical luminosity is also dependent on the neutron star mass and magnetic field strength, one might be tempted to conclude that the Vela pulsar is less massive and/or has a smaller magnetic field than the Crab pulsar. However, such a conclusion would conflict with the masses derived from observations of period discontinuities in these two pulsars (see p. 194) and with the field strengths derived from their rates of period increase (see p. 177).

In Sturrock's model the X-ray pulses from the Crab pulsar are generated by the synchrotron process. Consequently, a prediction of this model is that there should be a spectral discontinuity between the optical and X-ray regions. Furthermore, the optical emission should be polarized in the plane containing the field-line whose tangent is in the line of sight (as for curvature radio radiation), but the X-ray polarization should be perpendicular to this plane. Future observations will undoubtedly test these predictions. The observed form of the optical position angle and variations in fractional polarization are consistent with emission from the vicinity of a magnetic pole with the peak of the pulse originating very close to the magnetic axis. Provided the instantaneous emission region is of finite size, low polarization would be expected near the pulse peak because of the rapid variation in projected field angles near the axis.

In a different interpretation of the polarization of the Crab optical pulses, Ferguson (1973) and Cocke, Ferguson, and Muncaster (1973) have proposed that the degree of polarization is proportional to the projected length of a vector fixed in the rotating star, with the position angle of the observed polarization at a fixed angle to the projected direction of this vector. Observations by these authors suggest that the fractional polarization reaches a maximum value of about 10 percent about 2.5 ms (30° of longitude) before the pulse peak and the associated polarization minimum. On the basis of their assumption, the degree of polarization is maximum when the vector is perpendicular to the line of sight, and minimum when the two are most closely aligned. The observed points of maximum and minimum polarization are separated by less than 90°, so relativistic compression is invoked to reduce the apparent longitude range. Good fits to the curves of observed position angles and fractional polarization are obtained with $\gamma_\phi \approx 1.21$ for the main pulse and $\gamma_\phi \approx 1.10$ for the interpulse. In this model, unlike those of Zheleznyakov and Shaposhnikov (1972) and Smith (1973), the pulse profile is determined by an intrinsic beaming mechanism and is only slightly modified by the co-rotation beaming.

As described in Chapter 5 (p. 78), pulsed optical and γ-ray emission has recently been observed from the Vela pulsar. In contrast to the Crab pulsar, the pulse shapes in each of the three frequency regimes are quite different. Before the discovery of the optical pulses, Thompson (1975) suggested that the γ-ray pulses are synchrotron emission from the open field-lines in the vicinity of the light cylinder. Because of the azimuthal deformation of field-lines near the light cylinder, the γ-ray pulse would lag behind the radio pulse (assumed to be emitted from the open field-lines closer to the star), as observed. The open field-lines subtend large angles near the light cylinder, so the γ-ray pulse width is greater. The larger γ-ray pulse width also allows radiation from the other pole to be seen, giving the double-pulse structure. However, as mentioned above (p. 201) the smaller spacing and symmetric location of the optical pulses with respect to the γ-ray pulses suggest that only one pole is involved. If this is the case, the radio pulse would be emitted in a direction perpendicular to the magnetic axis. It is interesting that the Crab pulse profiles can be interpreted in the same way, with the radio precursor component (see Figure 4-6, p. 70) being analogous to the Vela radio pulse.

The total luminosity of the Vela pulsar is dominated by the γ-ray component, and is about 0.5 percent of the rotational energy loss rate. No models have as yet been proposed for the γ-ray emission seen from PSR 1747−46 and PSR 1818−04. For these pulsars the γ-ray luminosity is apparently a much larger fraction of the total energy loss rate.

Subpulse Drift and Pulsar Turnoff

One of the more curious pulsar characteristics is the systematic drift of subpulses across the integrated profile observed in some pulsars (see Figure 3-6, p. 41). Several possible mechanisms for this modulation have been proposed. An absorbing screen located in the vicinity of the light cylinder and rotating differentially with respect to the star was suggested by Sutton and co-workers (1970). As mentioned above, cyclotron absorption may be significant if the wave frequency equals the electron gyro-frequency within the light cylinder. Problems with this model are as follows. Some pulsars have extremely regular drifting behavior—the P_3 modulation of PSR 0809+74 has a Q of at least 50—and such high stability is difficult to explain. Subpulses in the sources with stable drifting move from the trailing edge of the profile to the leading edge, implying that the screen is effectively rotating faster than the neutron star; this would be impossible if the screen were located near the light cylinder. Finally, this model does not account for the observed frequency dependence of subpulse separation.

Within the context of the relativistic beaming model (p. 213), Zhelez-nyakov (1971) proposed that the drifting subpulses result from rotation of a multilobed radiation pattern. With a multilobed pattern in which the lobes are equally spaced in the source frame and rotate uniformly about an axis parallel to the stellar rotation axis, one or two subpulses would be seen within a given pulse. The drift rate near the wings of the profile would be greater than near the center; such a variation in drift rate is similar to that seen for PSR 1919+21.

A similar model, which does not invoke relativistic compression to reduce the subpulse separation, has been proposed by Oster (1975). This model requires a fan-beam radiation pattern with P/P_2 (approximately 15 to 100) lobes rotating uniformly about a perpendicular axis, presumably the magnetic axis. The subpulse drift rate is then related directly to the rate of rotation of the radiation pattern about its axis. Changes in drift rate (and in P_3) can be achieved by varying the rotation rate of the pattern while keeping the lobe spacing (P_2) constant. A modification of this model, in which the rotating-beam pattern forms a two-dimensional grid, was used by Oster and Sieber (1976) to compute subpulse patterns for different combinations of lobe separation, tilt of the beam rotation axis, observer orientation, etc. Depending on these various angles, either regular subpulse drift or the more disorganized drift seen in some pulsars can be simulated.

Apart from the difficulty of accounting for such a rotating fan-beam in a physical system, the principal problem with these models is that the appearance of subpulse drift is dependent on the relative orientation of the observer and the star. As was pointed out earlier in this chapter (p. 205), the strong correlation of drifting properties with period derivative implies that orientation with respect to the observer is not the dominant factor. The observed frequency dependence of P_2 is also difficult to account for, as it implies a change with frequency in the spacing, and hence number, of lobes. In the Oster–Sieber model the integrated profile is determined by some modulation imposed on the underlying rotating-beam system. Why the separation of integrated profile components should have the same $\omega^{-1/4}$ frequency dependence as the lobe separation is not obvious.

In the Ruderman–Sutherland model for pulse emission (p. 206) subpulse drift arises from the motion of the sparking regions around the magnetic pole. (Exact co-rotation with the star is not enforced above the vacuum gap because $\mathbf{E} \cdot \mathbf{B}$ is nonzero in this region.) The period associated with this drift is

$$P_{\mathrm{d}} \approx 5.6 B_{12} P^{-2} \quad \text{(s)}, \tag{10-40}$$

where B_{12} is the polar magnetic field strength (B_0) in units of 10^{12} G. The subpulse drift rates corresponding to this motion are of the same order as the observed values. If the separation of sparks in the gap region is comparable to the gap height (about 10^3 cm), the predicted values of the subpulse separation, P_2, are about equal to those observed. Since both profile components and subpulses arise from regions of enhanced particle emission on the stellar surface, the frequency dependence of P_2 is automatically the same as that of component separation in this model. However, the direction of subpulse drift is dependent on observer orientation with respect to the magnetic axis, because sparks always drift about the magnetic axis in the same direction as the stellar rotation. A possible way of overcoming this orientation dependence is as follows: when the angle between the magnetic and rotation axes (α) is significantly different from zero, spark trajectories may be deformed in such a way that they reach the edge of the emission zone only on the equatorial side (Ruderman, 1976b). Subpulses in type SD pulsars would then always drift toward the leading edge of the profile, as observed.

Periodic fluctuations of pulse intensities can also be accommodated in the Ruderman–Sutherland model. These occur mainly in Type C pulsars and are often different for different components, but usually symmetric about the profile center (see Figure 3-4, p. 38). Since the line of sight passes close to the magnetic axis for Type C pulsars, the drift motion of sparks in these pulsars would result in a modulation of pulse intensities rather than drifting subpulses. Since the drift paths circle the magnetic axis, fluctuation characteristics would be symmetric about the pulse center, as observed.

One of the striking features of the plot of period derivative versus period (Figure 10-2) is that the type D pulsars are concentrated on the right, near the sloping line. If magnetic fields do not decay, and if the braking index is three, then pulsars evolve downwards and to the right along straight lines in this figure (see Figure 6-3, p. 110). The sloping line evidently represents an empirical maximum pulsar period. However, many pulsars must turn off before this period is reached, because, for a steady-state situation, the number of pulsars in equal logarithmic frequency intervals would be expected to be proportional to $P/\dot{P}$. The quantity $P/\dot{P}$, the characteristic time, is proportional to P^2 for $n = 3$. In fact, the distribution of the number of pulsars versus period (Figure 1-4, p. 9) has a peak near the center of the observed period range rather than at the long-period end. It is clear that some pulsars must cease to emit pulses at periods at least as short as 0.75 s.

As mentioned previously (p. 205), the sloping line in Figure 10-2 represents a constant value of the parameter $\dot{P}P^{-5}$. Evidently this parameter is related to the pulsar "age," in the sense that pulsars cease to emit pulsed radiation when $\dot{P}P^{-5}$ is small. From a study of the characteristics of pulse nulling (see Chapter 3, p. 36), Ritchings (1976) has shown that the fraction of time that a pulsar is in a null state is large when $\dot{P}P^{-5}$ is small (Figure 10-7). This result suggests that pulsars become old and eventually die (at least as far as radio observations are concerned) by spending an increasing fraction of time in a null state. Ritchings finds that larger null

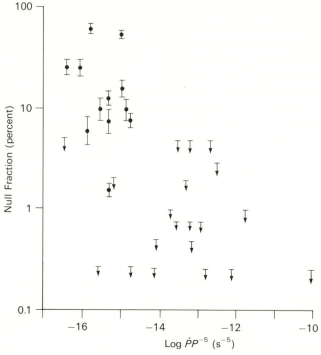

10-7 The fraction of time that a pulsar is in a null state plotted against the parameter $\dot{P}P^{-5}$ for 32 pulsars [Ritchings, 1976.]

fractions result from an increased amount of time being spent in the null state rather than shorter bursts of active pulsing.

Why do pulsars turn off? Since for a dipole field the surface magnetic field strength (B_0) is proportional to $(P\dot{P})^{1/2}$, the cutoff period represented by the sloping line in Figure 10-2 is proportional to $B_0^{1/3}$. Furthermore, since the field strength at the light cylinder varies as P^{-3}, the line of constant $\dot{P}P^{-5}$ represents a constant field strength at the light cylinder. Lyne, Ritchings, and Smith (1975) have pointed out that this fact could favor models in which the emission source is close to the light cylinder. Another possible explanation is that the cutoff is due to cyclotron absorption near the light cylinder. This occurs at a frequency

$$\omega \approx \frac{2\omega_B}{\gamma\theta^2}, \qquad (10\text{-}41)$$

where θ is the angle of propagation with respect to the field (Blandford and Scharlemann, 1976). As an example, if we take $\gamma \approx 10^3$ (as in the

Ruderman–Sutherland model), $\theta \approx 10^{-2}$ radians, and a pulsar with $P = 1$ s and $B_0 = 10^{12}$ G, the cutoff frequency near the light cylinder will be about 600 MHz. As the period increases this cutoff frequency moves steadily lower until the pulsar is no longer observable. There is some evidence that Types C and D pulsars have relatively steep high-frequency cutoffs at about these frequencies. Furthermore, the fractional circular polarization of integrated profiles is generally larger at high frequencies, and because charges of a given sign absorb only one sense of circular polarization, this also might be expected.

Within the Sturrock–Ruderman–Sutherland model, pair-production in the polar cap region is necessary for the production of pulses. As the star slows down, the potential drop through which the particles are accelerated (Eqn. 9-16) decreases as P^{-2}. Consequently, at a certain period the emitted γ-rays will no longer be sufficiently energetic to pair-produce and hence pulse emission will cease. Sturrock (1971) finds that the turnoff period is proportional to $B_0^{4/7}$: for a pulsar with $B_0 = 10^{12}$ G it would be about one second, as observed. Similar results are obtained by Ruderman and Sutherland (1975). The $B_0^{4/7}$ dependence is slightly stronger than the $B_0^{1/3}$ dependence observed; Ruderman and Sutherland point out that the predicted dependence is in better agreement with the observations if the braking index is 2.5 rather than 3.

In Chapter 8 (p. 164) it was suggested that decay of the pulsar magnetic field may account for the fact that the actual mean age of pulsars is much less than their mean characteristic age. Since most models for the pulse emission process rely on strong magnetic fields for the production of accelerating potentials, field decay would also result in turnoff of the pulsed emission. For example, in the magnetic pole models, the accelerating potential (Eqn. 9-16) is proportional to B_0. Field decay on a time scale of a few million years would therefore account for both the extremely large characteristic ages observed for some pulsars and the fact that the average active lifetime of pulsars is only a few million years.

Finally, another possible way in which pulsars could die would be alignment of the magnetic and rotational axes. If this process were important, it could account for the absence of interpulses in long-period pulsars. One would expect pulsars approaching turnoff to have wider-than-average profiles and braking indexes greater than three.

Appendix

TABLE OF PULSAR PARAMETERS

In this appendix we list the principal parameters for the 149 pulsars known at the time of writing (November 1976). Where known with sufficient accuracy, positions are quoted to the nearest second of time for right ascensions and arc second for declinations. Periods are given to four decimal digits, although in most cases they are known to much greater accuracy. Period derivatives are quoted in units of 10^{-15} s s^{-1} (1.0×10^{-15} s s^{-1} = 0.086 ns day^{-1}), and dispersion and rotation measures are given in their conventional units. Pulse equivalent widths, W_e, defined as pulse energy divided by peak flux density, are quoted in milliseconds. The mean flux density at 400 MHz, S_{400}, defined as pulse energy divided by the period, is in units of milliJanskys (1 mJy = 10^{-29} W m^{-2} Hz^{-1}). Except where marked with an asterisk, distances are derived from dispersion measures using Equation 8-1 with a mean electron density on the galactic plane $\langle n_e \rangle$ = 0.03 cm^{-3} and a scale height for the electron layer of 1000 pc. The effects of HII regions within 1 kpc of the sun were included following the methods of Prentice and ter Haar (1969). Distances marked with an asterisk are derived from either hydrogen-line absorption observations or pulsar–supernova remnant associations (Table 7-1, p. 127). Finally, the characteristic age, assuming a braking index $n = 3$, is given in units of millions of years. More accurate parameters, estimated uncertainties, and references may be obtained for most of the data in the table from the compilation of pulsar parameters by Taylor and Manchester (1975).

PSR	RA (1950)	Dec. (1950)	Galactic Long.	Lat.	Period, P (s)	Period derivative, $\dot{P}$ (10^{-15} ss^{-1})
0031−07	00^h 31^m 36^s	−07° 38′ 15″	110.4°	−69.8°	0.9429	0.40
0105+65	01 05	+65 50	124.6	3.3	1.2836	12.4
0138+59	01 38	+59 45	129.1	−2.3	1.2229	0.2
0153+61	01 53	+61 55	130.5	0.2	2.3516	189
0254−53	02 54 22	−53 14	269.8	−55.3	0.4477	
0301+19	03 01 42	+19 21 12	161.1	−33.3	1.3875	1.29
0329+54	03 29 11	+54 24 37	145.0	−1.2	0.7145	2.05
0355+54	03 55 00	+54 05	148.2	0.8	0.1563	4.39
0450−18	04 50 21	−18 04 21	217.1	−34.1	0.5489	5.78
0525+21	05 25 52	+21 58 18	183.8	−6.9	3.7454	40.06
0531+21	05 31 31	+21 58 54	184.6	−5.8	0.0331	422.69
0540+23	05 40 07	+23 28 15	184.4	−3.3	0.2459	15.43
0611+22	06 11 16	+22 31 42	188.8	2.4	0.3349	59.73
0628−28	06 28 51	−28 34 08	237.0	−16.8	1.2444	2.51
0736−40	07 36 51	−40 35 47	254.2	−9.2	0.3749	1.62
0740−28	07 40 47	−28 15 34	243.8	−2.4	0.1667	16.80
0809+74	08 09 03	+74 38 12	140.0	31.6	1.2922	0.16
0818−13	08 18 06	−13 41 23	235.9	12.6	1.2381	2.10
0823+26	08 23 50	+26 47 08	197.0	31.7	0.5306	1.67
0833−45	08 33 39	−45 00 10	263.6	−2.8	0.0892	125.03
0834+06	08 34 26	+06 20 43	219.7	26.3	1.2737	6.79
0835−41	08 35 34	−41 24 42	260.9	−0.3	0.7516	
0904+77	09 04	+77 40	135.3	33.7	1.5790	
0940−55	09 40 40	−55 40	278.6	−2.2	0.6643	
0943+10	09 43 27	+10 05 50	225.4	43.1	1.0977	3.53
0950+08	09 50 31	+08 09 44	228.9	43.7	0.2530	0.23
0959−54	09 59 52	−54 55	280.3	0.1	1.4365	
1055−52	10 55 49	−52 10 46	286.0	6.7	0.1971	
1112+50	11 12 49	+50 46 30	154.4	60.4	1.6564	2.6
1133+16	11 33 27	+16 07 35	241.9	69.2	1.1879	3.73
1154−62	11 54 45	−62 08 23	296.7	−0.2	0.4005	
1221−63	12 21 40	−63 52	300.0	−1.4	0.2164	
1237+25	12 37 12	+25 10 17	252.5	86.5	1.3824	0.95
1240−64	12 40 21	−64 06 51	302.1	−1.5	0.3884	
1323−62	13 24 00	−62 08	307.1	0.2	0.5299	

Dispersion measure (cm^{-3} pc)	Rotation measure (rad m^{-2})	Effective pulse width (ms)	Flux density at 400 MHz (mJy)	Distance (kpc)	$\frac{1}{2}P/\dot{P}$ (10^6 yr)
10.8	+10.0	42	27	0.4	37
30.1	−29	25	25	1.0	1.6
34.8	−50	35		3.0*	97
60		35	13	2.0	0.20
18		10	110	0.8	
15.6	−8.3	27	14	0.6	17.0
26.7	−63.7	8.7	2270	2.6*	5.5
57.0		3.5	56	1.5*	0.56
39.9	+15.0	19	130	2.4	1.5
50.9	−39.6	75	93	1.9	1.5
56.7	−42.3	1.9	480	2.0*	0.0012
77.5	+7	7.8	30	2.8	0.25
96.7	+69	7.1	48	3.5	0.089
34.3	+44.2	57	260	1.4	7.9
160.8	+21	18	110	2.0*	3.7
73.7	+150.0	8	360	2.0*	0.016
5.7	−11.7	45	81	0.2	128
40.9	−2.8	20	48	1.6	9.3
19.4	+5.9	6.0	41	0.8	5.0
69.0	+33.6	1.71	2800	0.5*	0.011
12.8	+24.5	17	79	0.5	3.0
147.6	+137	8	80	2.4*	
		80			
179.2	−30	13	150	0.5	
15.3		29	15	0.6	4.9
3.0	+1.8	9.5	260	0.1	17.3
130.6		50	56	4.4	
23.8	+44		100	0.8	
9.2	+3.2	14	5	0.4	10
4.8	+3.9	18	100	0.2	5.0
324.2	+509	17	110	11*	
98	−20			1.1	
9.3	−0.6	25	40	0.4	23
297.4	+168	14	64	14*	
318.4		18	190	8.0*	

* For explanation of asterisk see the remarks on measurement of distance in the introduction to the appendix.

PSR	RA (1950)	Dec. (1950)	Galactic		Period, P (s)	Period derivative, $\dot{P}$ (10^{-15} ss^{-1})
			Long.	Lat.		
1353−62	13^{h} 53^{m} 50^{s}	−62° 13′	310.5°	−0.6°	0.4557	
1359−50	13 59 43	−50	314.5	11.0	0.690	
1426−66	14 26 34	−66 10 06	312.6	−5.4	0.7854	
1449−64	14 49 26	−64 01 01	315.7	−4.4	0.1794	
1451−68	14 51 29	−68 31 31	313.9	−8.5	0.2633	0.09
1508+55	15 08 04	+55 42 56	91.3	52.3	0.7396	5.03
1530−53	15 30 22	−53 24 17	325.7	1.9	1.3688	
1541+09	15 41 14	+09 38 43	17.8	45.8	0.7484	0.41
1556−44	15 56 11	−44 30 16	334.5	6.4	0.2570	1.02
1557−50	15 57 09	−50 35 56	330.7	1.6	0.1925	5.06
1558−50	15 58 36	−50 53	330.7	1.3	0.8641	
1601−52	16 01 58	−52 56	329.7	−0.6	0.6759	
1604−00	16 04 38	−00 24 41	10.7	35.5	0.4218	0.30
1641−45	16 41 10	−45 53 37	339.2	−0.2	0.4550	20.10
1642−03	16 42 25	−03 12 31	14.1	26.1	0.3876	1.78
1700−32	17 00 10	−32 40	351.8	5.4	1.2117	0.7
1700−18	17 00 56	−18	3.8	14.0	0.802	
1706−16	17 06 33	−16 37 12	5.8	13.7	0.6530	6.36
1717−29	17 17 10	−29 32	356.5	4.3	0.6204	0.8
1718−32	17 18 40	−32 05	354.5	2.5	0.4771	0.7
1727−47	17 27 55	−47 42 22	342.6	−7.7	0.8296	180
1730−22	17 30 15	−22 29	4.0	5.8	0.8716	0.0
1730−07	17 30 51	−07 18	17.1	13.7	0.4193	
1742−30	17 42 35	−30 38	358.6	−0.9	0.3674	10.7
1747−46	17 47 58	−46 56 42	345.0	−10.2	0.7423	70.
1749−28	17 49 49	−28 06 00	1.5	−1.0	0.5625	8.15
1754−24	17 54 37	−24 21	5.3	0.0	0.2340	
1813−26	18 13 10	−26 50	5.2	−4.8	0.5928	
1818−04	18 18 14	−04 29 03	25.5	4.7	0.5980	6.32
1819−22	18 19 50	−22 53	9.4	−4.3	1.8742	0.6
1822−09	18 22 40	−09 36	21.5	1.4	0.7689	
1826−17	18 26 55	−17 54	14.6	−3.4	0.3071	5.59
1831−03	18 31 00	−03 40	27.7	2.3	0.6866	41.5
1831−04	18 31 40	−04 31	27.0	1.8	0.2901	0.10
1845−04	18 45 10	−04 05 32	28.9	−1.0	0.5977	51.9

Dispersion measure (cm^{-3} pc)	Rotation measure (rad m^{-2})	Effective pulse width (ms)	Flux density at 400 MHz (mJy)	Distance (kpc)	$\frac{1}{2}P/\dot{P}$ (10^6 yr)
434				15	
20		20	72	0.7	
65.3	−12	10	50	2.4	
70.7	−28	5	220	2.6	
8.6	+3	13	840	0.3	44
19.5	+0.8	13	47	0.9	2.3
24.8		25	73	0.8	
34.9		44	53	2.5	29
58.8	−2		190	2.2	4.0
270	+160	10		9.0*	0.60
169.5	+70		35	6.0	
30			59	1.0	
10.7	+6.5	8	24	0.4	22
450	−614	9	1300	4.9*	0.36
35.7	+16.5	4.0	350	0.2*	3.4
105.4	−25	45	92	0.5	27
40			100	0.2	
24.8	−2.5	11	69	0.2	1.6
45		30	30	1.6	12
126.3		16	95	4.7	11
121.9	−430	20	220	5.9	0.073
45		15	31	1.6	
70				3.4	
8	+80	8	54	2.9	0.54
20.7	+18	20	120	0.7	0.17
50.8	+95.0	7	1070	1.0*	1.1
188				6.3	
90		45	26	3.5	
84.3	+70.5	11	150	1.5*	1.5
140		70	22	5.7	49
19.3	+69.0	11	25	0.6	
207		18	85	8.9	0.87
235		20	61	9.4	0.26
68		25	81	2.3	46
141.9		20	55	4.9	0.18

* For explanation of asterisk see the remarks on measurement of distance in the introduction to the appendix.

PSR	RA (1950)	Dec. (1950)	Galactic Long.	Galactic Lat.	Period, P (s)	Period derivative, $\dot{P}$ $(10^{-15}\,ss^{-1})$
1845−01	18^h 45^m 50^s	−01° 27′ 30″	31.3°	0.0°	0.6594	5.2
1846−06	18 46 07	−06 43	26.7	−2.4	1.4512	45.7
1857−26	18 57 43	−26 04 54	10.3	−13.5	0.6122	0.16
1859+03	18 59 02	+03 26 46	37.2	−0.6	0.6554	7.50
1900+05	19 00 15	+05 52	39.5	0.2	0.7465	
1900−06	19 00 30	−06 35	28.5	−5.6	0.4318	3.4
1900+01	19 00 58	+01 31 09	35.7	−1.9	0.7293	4.03
1901+10	19 01 40	+10 00	43.3	1.8	1.8565	
1904+12	19 04 57	+12 40	46.1	2.3	0.8270	
1906+09	19 06 40	+09 10	43.2	0.3	0.8302	
1907+00	19 07 01	+00 03 03	35.1	−4.0	1.0169	5.51
1907+02	19 07 08	+02 49 56	37.6	−2.7	0.4949	2.76
1907+10	19 07 27	+10 57 08	44.8	1.0	0.2836	2.63
1908+12	19 08 06	+12 28	46.2	1.6	1.4417	
1910+20	19 10 34	+20 59 26	54.1	5.0	2.2329	10.18
1910+10	19 10 30	+10 30	44.8	0.1	0.4093	
1911+13	19 11 10	+13 54	47.9	1.6	0.5214	
1911−04	19 11 15	−04 45 59	31.3	−7.1	0.8259	4.06
1911+09	19 11 30	+09 30	44.0	−0.6	1.2419	
1911+03	19 11 48	+03 54	39.1	−3.2	2.3303	
1911+11	19 11 57	+11 15	45.6	0.2	0.6009	
1913+10	19 13 05	+10 05	44.7	−0.6	0.4045	
1913+167	19 13 10	+16 40	50.5	2.4	1.6162	
1913+16	19 13 13	+16 00 24	50.0	2.1	0.0590	0.0088
1914+09	19 14 10	+09 47	44.6	−1.0	0.2702	
1914+13	19 14 40	+13 06	47.6	0.4	0.2818	
1915+13	19 15 22	+13 48 28	48.3	0.6	0.1946	7.20
1916+14	19 16 00	+14 40	49.1	0.9	1.1808	
1917+00	19 17 17	+00 16 03	36.5	−6.2	1.2722	7.67
1918+19	19 18 53	+19 43 02	53.9	2.6	0.8210	0.89
1919+14	19 19 10	+14 12	49.0	−0.0	0.6181	
1919+21	19 19 36	+21 47 16	55.8	3.5	1.3373	1.34
1919+20	19 19 40	+20 00	54.2	2.6	0.7606	
1920+20	19 20 08	+20 10	54.4	2.6	1.1727	
1920+21	19 20 42	+21 05 13	55.3	2.9	1.0779	8.2

Dispersion measure (cm^{-3} pc)	Rotation measure (rad m^{-2})	Effective pulse width (ms)	Flux density at 400 MHz (mJy)	Distance (kpc)	$\frac{1}{2}P/\dot{P}$ (10^6 yr)
63	+ 575	55	57	5.4	2.0
152		35	31	5.7	0.50
38.1	− 13	35	82	1.5	61
402.9	− 238	35	74	20*	1.4
170		30	25	5.7	
205		12	16	11.2	2.0
243.4		18	103	9.5	2.9
140		50	2	5.0	
260		30	2	11	
250		45	5	8.5	
111		8	26	4.3	2.9
166		14	15	6.4	2.8
144		5.6	61	5.0	1.7
260		40	5	9.9	
84		13	10	3.2	3.5
140		25	2	4.7	
145		25	6	5.2	
89.4		7.5	71	3.7	3.2
155		45	3	5.3	
35		480	43	1.2	
80		20	5	2.7	
240		40	30	8.4	
55		45	6	1.9	
167		10	12	6.2	106
60		20	45	2.0	
230		20	25	7.9	
94	+ 264	4.5	62	3.2	0.43
30		25	6	1.0	
85		18	34	3.4	2.6
140		44	38	5.3	14.6
95		30	17	3.2	
12.4	− 18.2	25	56	0.4	15
70		30	1	2.5	
215		55	6	8.7	
220		15	48	9.2	2.1

* For explanation of asterisk see the remarks on measurement of distance in the introduction to the appendix.

PSR	RA (1950)	Dec. (1950)	Galactic Long.	Galactic Lat.	Period, P (s)	Period derivative, $\dot{P}$ $(10^{-15}\,\mathrm{ss^{-1}})$
1921+17	19ʰ 21ᵐ 06ˢ	+17° 00′	51.7°	0.9°	0.5472	
1922+20	19 22 30	+20 30	55.0	2.3	0.2377	
1924+19	19 24 16	+19 20	54.1	1.4	1.3460	
1924+16	19 24 30	+16 42	51.9	0.1	0.5798	
1924+14	19 24 35	+14 30	49.9	−1.0	1.3249	
1925+18	19 25 00	+18 50	53.8	1.0	0.4827	
1925+22	19 25 08	+22 30	57.0	2.7	1.4310	
1925+188	19 25 27	+18 50	53.8	0.9	0.2983	
1927+18	19 27 18	+18 40	53.9	0.4	1.2204	
1927+13	19 27 35	+13 10	49.1	−2.3	0.7600	
1929+15	19 29 30	+15 30	51.4	−1.6	0.3143	
1929+10	19 29 52	+10 53 03	47.4	−3.9	0.2265	1.16
1930+20	19 30 00	+20 15	55.6	0.6	0.2682	
1930+22	19 30 30	+22 15	57.4	1.5	0.1444	63
1930+13	19 30 58	+13 00	49.4	−3.1	0.9283	
1933+17	19 33 15	+17 40	53.7	−1.3	0.6544	
1933+16	19 33 32	+16 09 58	52.4	−2.1	0.3587	6.00
1933+15	19 33 40	+15 30	51.9	−2.4	0.9673	
1939+17	19 39 47	+17 40	54.5	−2.7	0.6962	
1942+17	19 42 15	+17 50	54.9	−3.1	1.9968	
1943+18	19 43 18	+18 30	55.6	−3.0	1.0687	
1944+22	19 44 16	+22 40	59.3	−1.0	1.3344	
1944+17	19 44 39	+17 58 15	55.3	−3.5	0.4406	0.02
1946+35	19 46 34	+35 32 38	70.7	5.0	0.7173	7.05
1952+29	19 52 22	+29 15 22	65.9	0.8	0.4266	0.0019
2002+31	20 02 54	+31 28 34	69.0	0.0	2.1112	74.58
2016+28	20 16 00	+28 30 30	68.1	−4.0	0.5579	0.14
2020+28	20 20 33	+28 44 43	68.9	−4.7	0.3434	1.89
2021+51	20 21 25	+51 45 08	87.9	8.4	0.5291	3.04
2024+21	20 24 55	+21 40	63.6	−9.5	0.3981	
2028+22	20 28 24	+22 20	64.6	−9.8	0.6305	
2045−16	20 45 47	−16 27 48	30.5	−33.1	1.9615	10.96
2106+44	21 06 30	+44 30	86.9	−2.0	0.4148	0.1
2111+46	21 11 38	+46 31 42	89.0	−1.3	1.0146	0.71
2148+63	21 48 40	+63 15	104.3	7.4	0.3801	0.16

Dispersion measure (cm^{-3} pc)	Rotation measure (rad m^{-2})	Effective pulse width (ms)	Flux density at 400 MHz (mJy)	Distance (kpc)	$\frac{1}{2}P/\dot{P}$ (10^6 yr)
135		25	2	4.7	
215		20	4	8.4	
420		55	2	17	
170		25	10	5.7	
205		45	9	7.3	
250		35	3	9.0	
180		60	6	7.1	
90		30	3	3.1	
110		30	3	3.7	
200		25	5	7.8	
120		30	3	4.2	
3.17	−8.6	5.5	610	0.1	3.1
200		20	7	6.9	
219		8	11	8.1	0.036
165		35	2	6.5	
210		35	3	7.6	
158.5	−1.9	6.5	360	6.0*	0.95
165		30	2	6.3	
175		65	3	6.8	
160		45	1	6.3	
240		35	3	10	
140		30	3	4.9	
16.3		23	33	0.6	286
129.1		21	120	8.5*	1.6
7.91		13	20	0.7	3600
233		15	14	7.8	0.45
14.1	−34.6	14	290	1.0*	59
24.6	−74.7	6.7	250	2.0*	2.9
22.5	−6.5	11.5	310	0.8	2.7
90		20	3	4.1	
60		25	9	2.4	
11.5	−10.8	42	61	0.4	2.8
129		30	38	0.8	66
141.5	−223.7	29	190	4.0*	22
125		15	25	0.8	38

* For explanation of asterisk see the remarks on measurement of distance in the introduction to the appendix.

PSR	RA (1950)	Dec. (1950)	Galactic		Period, P (s)	Period derivative, $\dot{P}$ (10^{-15} ss^{-1})
			Long.	Lat.		
2154 + 40	21$^{\rm h}$ 54$^{\rm m}$ 57$^{\rm s}$	+40° 03′ 30″	90.5°	−11.3°	1.5252	2.9
2217 + 47	22 17 46	+47 39 48	98.4	−7.6	0.5384	2.76
2223 + 65	22 23 30	+65 22	108.6	6.9	0.6825	9.5
2255 + 58	22 55 46	+58 54 30	108.8	−0.5	0.3682	
2303 + 30	23 03 34	+30 43 49	97.7	−26.7	1.5758	2.91
2305 + 55	23 05	+55 26	108.6	−4.2	0.4750	0.1
2319 + 60	23 19 41	+60 08 00	112.1	−0.6	2.2564	7.6
2324 + 60	23 24	+60 55	112.9	−0.0	0.2336	0.36
2327 − 20	23 27 50	−20 18	49.6	−70.2	1.6436	

Dispersion measure (cm^{-3} pc)	Rotation measure (rad m^{-2})	Effective pulse width (ms)	Flux density at 400 MHz (mJy)	Distance (kpc)	$\frac{1}{2}P/\dot{P}$ (10^6 yr)
71.0	−44.0	52	39	3.2	8.3
43.5	−35.3	7.3	63	1.6	3.1
35		34	38	0.7	1.1
148		13	42	2.2	
49.9		17	25	3.1	8.6
45		25	23	1.6	75
96	−224	65	70	2.8*	4.7
120		14	41	0.8	10
18				0.9	

* For explanation of asterisk see the remarks on measurement of distance in the introduction to the appendix.

Symbols

We have endeavored, insofar as possible, to avoid using the same symbol for more than one quantity. Nevertheless, some multiple use has proved inevitable, and the following glossary will be useful in resolving any ambiguities that may arise. Dummy mathematical variables and symbols used only once are not listed. Boldface symbols denote vector quantities.

P_3	Spacing, in periods, between successive bands of drifting subpulses	41
$P(r)$	Pressure at distance r from the center of a star	172
q	Spatial frequency of density irregularities	140
Q	Fraction of a pulsar spin-up that decays away	115
r	Radial distance from the center of a star	66
r_p	Linear scale of scattering-induced diffraction pattern at the earth	140
$\mathbf{r}_s$	Vector from solar system barycenter to observing site	103
R	Radius of a star	86
R	Rate of occurrence of small, random period irregularities	117
R	Distance from the galactic center	153
R_p	Radius of polar cap region	180
R_L	Radius of the velocity-of-light cylinder	179
RM	Rotation measure	134
$\mathscr{R}$	Pulsar timing residual, i.e., observed minus predicted time of arrival	105
S	Flux density	8
S_0	Minimum flux density of a pulsar survey	158
S_{400}	Flux density at 400 MHz	158
t, t_b	Pulse arrival time at solar system barycenter	103
t_q	Expected time interval between starquake events	194
t_s	Pulse arrival time at the observing site	103
T	Temperature	
T	Characteristic time, $P/\dot{P}$	9
T_b	Brightness temperature	7
T_r	Receiver noise temperature	158
T_s	Spin or excitation temperature of a gas	124
T_0	Time of periastron passage in a binary orbit	93
U_p	Energy density of a plasma	209
U_A	Energy density of Alfvén waves	210
U_B	Energy density of magnetic field	216
U_L	Energy density of longitudinal waves	210
U_R	Energy density of radiation	209
v	Velocity	67

Literature Cited

Ables, J. G., and R. N. Manchester. 1976. *Astron. Astrophys.*, **50**:177.

Aitken, D. K., and P. G. Polden. 1971. *Nature Phys. Sci.*, **233**:45.

Albats, P., G. M. Frye, G. B. Thomson, V. D. Hopper, O. B. Mace, J. A. Thomas, and J. A. Staib. 1974. *Nature*, **251**:400.

Anderson, B., A. G. Lyne, and R. J. Peckham. 1975. *Nature*, **258**:215

Anderson, P. W., and N. Itoh. 1975. *Nature*, **256**:25.

Andrew, B., H. Branson, and D. Wills. 1964. *Nature*, **203**:171.

Argyle, E., and J. F. R. Gower. 1972. *Astrophys. J.* (*Letters*), **175**:L89.

Ash, M. E., I. I. Shapiro, and W. B. Smith. 1967. *Astron. J.*, **72**:338.

Avni, Y., and J. N. Bahcall. 1975. *Astrophys. J.*, **197**:675.

Baade, W. 1942. *Astrophys. J.*, **96**:188.

Baade, W., and F. Zwicky. 1934. *Proc. Nat. Acad. Sci.*, **20**:254.

Backer, D. C. 1970a. *Nature*, **228**:1297.

———. 1970b. *Nature*, **228**:42.

———. 1973. *Astrophys. J.*, **182**:245.

———. 1974. *Astrophys. J.*, **190**:667.

———. 1975. *Astron. Astrophys.*, **43**:395.

———. 1976. *Astrophys. J.*, **209**:895.

Backer, D. C., V. Boriakoff, and R. N. Manchester. 1973. *Nature Phys. Sci.*, **243**:77.

Backer, D. C., and J. R. Fisher. 1974. *Astrophys. J.*, **189**:137.

Backer, D. C., J. M. Rankin, and D. B. Campbell. 1975. *Astrophys. J.*, **197**:481.

———. 1976. *Nature*, **263**:202.

Backer, D. C., and R. A. Sramek. 1976. *Astron. J.*, **81**:1430.

Baym, G., and C. Pethick. 1975. *Ann. Rev. Nuclear Sci.*, **25**:27.

Baym, G., C. Pethick, D. Pines, and M. Ruderman. 1969. *Nature*, **224**:872.

Baym, G., and D. Pines. 1971. *Ann Phys.*, **66**:816.

Becklin, E. E., and D. E. Kleinmann. 1968. *Astrophys. J. (Letters)*, **152**:L25.

Becklin, E. E., J. Kristian, K. Matthews, and G. Neugebauer. 1973. *Astrophys. J. (Letters)*, **186**:L137.

Bell, S. J., and A. Hewish. 1967. *Nature*, **213**:1214.

Biermann, P., and B. M. Tinsley. 1974. *Astron Astrophys.*, **30**:1.

Blandford, R. D., and E. T. Scharlemann. 1976. *Mon. Not. Roy. Astron. Soc.*, **174**:59.

Blanford, R. D., and S. A. Teukolsky. 1976. *Astrophys. J.*, **205**:580.

Boksenberg, A., B. Kirkham, W. A. Towlson, T. E. Venis, B. Bates, G. R. Courts, and P. P. D. Carson. 1972. *Nature Phys. Sci.*, **240**:127.

Bolton, J. G., G. Stanley, and O. B. Slee. 1949. *Nature*, **164**:101.

Booth, R. S., and A. G. Lyne. 1976. *Mon. Not. Roy. Astron. Soc.*, **174**:53P.

Bowyer, S., E. Byram, T. Chubb, and H. Friedmann. 1964. *Science*, **146**:912.

Boynton, P. E., E. J. Groth, D. P. Hutchinson, G. P. Nanos, R. B. Partridge, and D. T. Wilkinson. 1972. *Astrophys. J.*, **175**:217.

Boynton, P. E., E. J. Groth, R. B. Partridge, and D. T. Wilkinson. 1969. *Astrophys. J. (Letters)*, **157**:L197.

Bridle, A. H. 1970. *Nature*, **225**:1035.

Bridle, A. H., and V. R. Venugopal. 1969. *Nature*, **224**:545.

Buccheri, R. 1976. The time structure of gamma-ray emission from the Crab and Vela pulsars. In *The Structure and Content of the Galaxy and Galactic Gamma Rays*, C. E. Fichtel and F. M. Stecker, eds. Greenbelt, Md.: Goddard Space Flight Center, p. 52.

Budden, K. G. 1961. *Radio Waves in the Ionosphere*, Cambridge University Press, p. 432.

Cameron, A. G. W. 1970. *Ann. Rev. Astron. Astrophys.*, **8**:179.

Canuto, V. 1974. *Ann. Rev. Astron. Astrophys.*, **12**:167.

———. 1975. *Ann. Rev. Astron. Astrophys.*, **13**: 335.

Canuto, V., and S. M. Chitre. 1973. *Nature Phys. Sci.*, **243**:63.

Chanmugam, G. 1973. *Astrophys. J. (Letters)*, **182**:L39.

Chau, W. Y., R. N., Henriksen, and D. R. Rayburn. 1971. *Astrophys. J. (Letters)*, **168**:L79.

Cheng, A., M. Ruderman, and P. Sutherland. 1976. *Astrophys. J.*, **203**:209.

Chiu, H. Y., and V. Canuto. 1971. *Astrophys. J.*, **163**:577.

Clark, G. W. 1975. Enrico Fermi Summer School on Physics and Astrophysics of Neutron Stars and Black Holes, Varenna, Italy.

Clark, G. W., H. V. Bradt, W. H. G. Lewin, T. H. Markart, H. W. Schnopper, and G. F. Sprott. 1973. *Astrophys. J.*, **179**:263.

Clark, D. H., and J. L. Caswell. 1976. *Mon. Not. Roy. Astron. Soc.* **174**:267.

Clemence, G. M., and V. Szebehely. 1967. *Astron. J.*, **72**:1324.

Cocke, W. J. 1973. *Astrophys. J.*, **184**:291.

———. 1975. *Astrophys. J.*, **202**:773.

Cocke, W. J., M. J. Disney, and D. J. Taylor. 1969. *Nature*, **221**:525.

Cocke, W. J., D. C. Ferguson, and G. W. Muncaster. 1973. *Astrophys. J.*, **183**:987.

Cocke, W. J., G. W. Muncaster, and T. Gehrels. 1971. *Astrophys. J. (Letters)*, **169**:L119.

Cocke, W. J., and A. G. Pacholczyk. 1976. *Astrophys. J. (Letters)*, **204**:L13.

Cohen, J. M., W. D. Langer, L. C. Rosen, and A. G. W. Cameron. 1970. *Astrophys. Space Sci.*, **6**:228.

Cohen, J. M., and E. T. Toton. 1971. *Astrophys. Letters*, **7**:213.

Cole, T. W. 1969. *Nature*, **221**:29.

Comella, J. M., H. D. Craft, R. V. E. Lovelace, J. M. Sutton, and G. L. Tyler. 1969. *Nature*, **221**:453.

Condon, J. J., and D. C. Backer. 1975. *Astrophys. J.*, **197**:31.

Cordes, J. M. 1975a. *Astrophys. J.*, **195**:193.

———. 1975b. *Astrophys. J.*, **208**:944.

———. 1975c. *Pulsar Microstructure: Time Scales, Spectra, Polarization, and Radiation Models*, thesis, University of California, San Diego.

Craft, H. D. 1970. *Radio Observations of the Pulse Profiles and Dispersion Measures of Twelve Pulsars*, Cornell Center for Radiophysics and Space Research, Report 395.

Craft, H. D., J. M. Comella, and F. D. Drake. 1968. *Nature*, **218**:1122.

Cronyn, W. M. 1970. *Science*, **168**:1453.

Davidsen, A., B. Margon, J. Liebert, H. Spinrad, J. Middleditch, G. Chanan, K. O. Mason, and P. W. Sanford. 1975. *Astrophys. J. (Letters)*, **200**:L19.

Davidson, K., and W. Tucker. 1970. *Astrophys. J.*, **161**:437.

Davidson, P. J. N., J. L. Culhane, and L. V. Morrison. 1975. *Nature*, **253**:610.

Davies, J. G., G. C. Hunt, and F. G. Smith. 1969. *Nature*, **221**:27.

Davies, J. G., A. G. Lyne, and J. H. Seiradakis. 1972. *Nature*, **240**:229.

———. 1973. *Nature Phys. Sci.*, **244**:84.

———. 1977. *Mon. Not. Roy. Astron. Soc.*, in press.

Downs, G. S., and P. E. Reichley. 1971. *Astrophys. J. (Letters)*, **163**:L11.

Drake, F. D., and H. D. Craft. 1968. *Nature*, **220**:231.

Dulk, G. A., and O. B. Slee. 1975. *Astrophys. J.*, **199**:61.

Duncan, J. C. 1939. *Astrophys. J.*, **89**:482.

Eardley, D. M. 1975. *Astrophys. J. (Letters)*, **196**:L59.

Eastlund, B. J. 1968, *Nature*, **220**:1293.

———. 1970. *Nature*, **225**:430.

Elitzur, M. 1974. *Astrophys. J.*, **190**:673.

Endean, V. G. 1972. *Mon. Not. Roy. Astron. Soc.*, **158**:13.

———. 1974. *Astrophys. J.*, **187**:359.

Epstein, R. I., and V. Petrosian. 1973. *Astrophys. J.*, **183**:611.

Esposito, L. W., and E. R. Harrison. 1975. *Astrophys. J. (Letters)*, **196**:L1.

Ewart, G. M., R. A. Guyer, and G. Greenstein. 1975. *Astrophys. J.*, **202**:238.

Ewing, M. S., R. A. Batchelor, R. D. Friefeld, R. M. Price, and D. H. Staelin. 1970 *Astrophys. J. (Letters)*, **162**:L169.

Fazio, G. G., H. F. Helmken, E. O'Mongain, and T. C. Weekes. 1972. *Astrophys. J. (Letters)*, **175**:L117.

Feibelman, P. J. 1971. *Phys. Rev. D.*, **4**:1589.

Ferguson, D. C. 1973. *Astrophys J.*, **183**:977.

Ferguson, D. C., W. J. Cocke, and T. Gehrels. 1974. *Astrophys J.*, **190**:375.

Ferguson, D. C., D. A. Graham, B. B. Jones, J. H. Seiradakis, and R. Wielebinski. 1976. *Nature*, **260**:25.

Forman, W., R. Giacconi, C. Jones, E. Schreier, and H. Tananbaum. 1974. *Astrophys. J. (Letters)*, **193**:L67.

Fritz, G., R. C. Henry, J. F. Meekins, T. A. Chubb, and H. Friedman. 1969. *Science*, **164**:709.

Galt, J. A., and A. G. Lyne. 1972. *Mon. Not. Roy. Astron. Soc.*, **158**:281.

Gardner, F. F., D. Morris, and J. B. Whiteoak. 1969. *Austral. J. Phys.*, **22**:813.

Gerola, H., M. Kafatos, and R. McCray. 1974. *Astrophys. J.*, **189**:55.

Giacconi, R. 1975a. *Ann. N.Y. Acad. Sci.*, **262**:312.

———. 1975b. *Astrophysics and Gravitation, Proc. 16th International Solvay Congress*, Brussels: L'Université de Bruxelles, p. 27.

Giacconi, R., S. Murray, H. Gursky, E. Kellogg, E. Schreier, T. Matilsky, D. Koch, and H. Tananbaum. 1974. *Astrophys. J. Suppl. Ser.*, **27**:37.

Ginzburg, V. L. 1970. *The Propagation of Electromagnetic Waves in Plasmas*, 2nd ed., New York: Pergamon.

Ginzburg. V. L., and S. I. Syrovatskii. 1965. *Ann. Rev. Astron. Astrophys.*, **3**:297.

Ginzburg, V. L. and V. V. Zheleznyakov. 1970. *Comm. Astrophys. Space Sci.*, **2**:167.

———. 1975. *Ann. Rev. Astron. Astrophys.*, **13**:511.

Ginzburg, V. L., V. V. Zheleznyakov, and V. V. Zaitsev. 1969. *Astrophys. Space Sci.*, **4**:464.

Gold, T. 1968. *Nature*, **218**:731.

———. 1969. *Nature*, **221**:25.

Goldreich, P. 1970. *Astrophys. J. (Letters)*, **160**:L11.

Goldreich, P., and W. H. Julian. 1969. *Astrophys. J.*, **157**:869.

Goldreich, P., and D. A. Keeley. 1971. *Astrophys. J.*, **170**:463.

Gómez-Gonzáles, J., and M. Guélin. 1974. *Astron. Astrophys.*, **32**:441.

Gordon, K. J., and C. P. Gordon. 1973. *Astron. Astrophys.*, **27**:119.

———. 1975. *Astron. Astrophys.*, **40**:27.

Goss, W. M., R. N. Manchester, W. B. McAdam, and R. H. Frater. 1977. *Mon. Not. Roy. Astron. Soc.*, in press.

Gott, J. R., J. E. Gunn, and J. P. Ostriker. 1970. *Astrophys. J. (Letters)*, **160**:L91.

Gould, R. J. 1965. *Phys. Rev. Letters*, **15**:577.

Gower, J. F. R., and E. Argyle. 1972. *Astrophys. J. (Letters)*, **171**:L23.

Graham, D. A., U. Mebold, K. H. Hesse, D. L. Hills, and R. Wielebinski. 1974. *Astron. Astrophys.*, **37**:405.

Greenstein, G. S. 1972. *Astrophys. J.*, **177**:251.

———. 1975. *Astrophys. J.*, **200**:281.

———. 1976. *Astrophys. J.*, **208**:836.

Greenstein, G. S., and A. G. W. Cameron. 1969. *Nature*, **222**:862.

Groth, E. J. 1975a. *Astrophys. J. Suppl. Ser.* **29**:431.

———. 1975b. *Astrophys. J.*, **200**:278.

Gunn, J. E., and J. P. Ostriker. 1970. *Astrophys. J.*, **160**:979.

Hamilton, P. A., P. M. McCulloch, R. N. Manchester. and J. G. Ables. 1977. *Mon. Not. Roy. Astron. Soc.*, in press.

Hankins, T. H. 1971. *Astrophys. J.*, **169**:487.

———. 1972. *Astrophys. J. (Letters)*, **177**:L11.

———. 1973. *Astrophys. J. (Letters)*, **181**:L49.

Hardee, P. E., and W. K. Rose. 1974. *Astrophys. J. (Letters)*, **194**:L35.

Hari-Dass, N. D., and V. Radhakrishnan. 1975. *Astrophys. Letters*, **16**:135.

Harnden, F. R., and P. Gorenstein. 1973. *Nature*, **241**:107.

Harrison, E. R., and E. Tademaru. 1975. *Astrophys. J.*, **201**:447.

Hegyi, D., R. Novick, and P. Thaddeus. 1971. *The Crab Nebula, I.A.U. Symposium No. 46*, Dordrecht: Reidel, p. 129.

Heiles, C., D. B. Campbell, and J. M. Rankin. 1970. *Nature*, **226**:529.

Heiles, C., and J. M. Rankin. 1971. *Nature Phys. Sci.*, **231**:97.

Helfand, D. J., R. N. Manchester, and J. H. Taylor. 1975. *Astrophys. J.*, **198**:661.

Henning, K., and H. J. Wendker. 1975. *Astron. Astrophys.*, **44**:91.

Henriksen, R. N., and J. A. Norton. 1975a. *Astrophys. J.*, **201**:431.

———. 1975b. *Astrophys. J.*, **201**:719.

———. 1976. Unpublished work.

Henriksen, R. N., and D. R. Rayburn. 1974. *Mon. Not. Roy. Astron. Soc.*, **166**:409.

Hensberge, G., E. P. J. van den Heuvel, and M. H. Paes De Barros. 1973. *Astron. Astrophys.*, **29**:69.

Hesse, K. H. 1973. *Astron. Astrophys.*, **27**:373.

Hesse, K. H., W. Sieber, and R. Wielebinski. 1973. *Nature Phys. Sci.*, **245**:57.

Hesse, K. H., and R. Wielebinski. 1974. *Astron. Astrophys.*, **31**:409.

Hewish, A. 1975. *Science*, **188**:1079.

Hewish, A., S. J. Bell, J. D. H. Pilkington, P. F. Scott, and R. A. Collins. 1968. *Nature*, **217**:709.

Hewish, A., and S. J. Burnell. 1970. *Mon. Not. Roy. Astron. Soc.*, **150**:141.

Hewish, A., and S. E. Okoye. 1964. *Nature*, **203**:171.

Hillier, R. R., W. R. Jackson, A. Murray, R. M. Redfern, and R. G. Sale. 1970. *Astrophys. J. (Letters)*, **162**:L177.

Hinata, S. 1976. *Astrophys. J.*, **206**:282.

Hinata, S., and E. A. Jackson. 1974. *Astrophys. J.*, **192**:703.

Hjellming, R. M., C. P. Gordon, and K. J. Gordon. 1969. *Astron. Astrophys.*, **2**:202.

Hoffman, B. 1968. *Nature*, **218**:757.

Horowitz, P., C. Papaliolios, and N. P. Carleton. 1971. *Astrophys. J.* (*Letters*), **163**:L5.

———. 1972. *Astrophys. J.* (*Letters*), **172**:L51.

Huguenin, G. R., J. H. Taylor, and D. J. Helfand. 1973. *Astrophys. J.* (*Letters*), **181**:L139.

Huguenin, G. R., J. H. Taylor, R. M. Hjellming, and C. M. Wade. 1971. *Nature Phys. Sci.*, **234**:50.

Huguenin, G. R., J. H. Taylor, and T. H. Troland. 1970. *Astrophys. J.*, **162**:727.

Hulse, R. A., and J. H. Taylor. 1974. *Astrophys. J.* (*Letters*), **191**:L59.

———. 1975a. *Astrophys. J.* (*Letters*), **195**:L51.

———. 1975b. *Astrophys. J.* (*Letters*), **201**:L55.

Ichimaru, S. 1970. *Nature*, **226**:731.

Illarionov, A. F., and R. A. Sunyaev. 1975. *Astron. Astrophys.* **39**:185.

Ilovaisky, S. A., and J. Lequeux. 1972. *Astron. Astrophys.*, **20**:347.

Jones, C., W. Forman, H. Tananbaum, E. Schreier, H. Gursky, E. Kellogg, and R. Giacconi. 1973. *Astrophys. J.* (*Letters*), **181**:L43.

Jones, C., and W. Liller. 1973. *Astrophys. J.* (*Letters*), **184**:L65.

Joss, P. C., and W. A. Fechner. 1975. *Ann. N.Y. Acad. Sci.*, **262**:385.

Julian, W. H. 1973. *Astrophys. J.*, **183**:967.

Kaplan, S. A., and V. N. Tsytovich. 1973a. *Nature Phys. Sci.*, **241**:122.

———. 1973b. *Plasma Astrophysics*, Oxford: Pergamon.

Kellermann, K. I., I. I. K. Pauliny-Toth, and P. J. S. Williams. 1969. *Astrophys. J.*, **157**:1.

Kniffen, D. A., R. C., Hartman, D. J. Thompson, G. F. Bignami, C. E. Fichtel, T. Tümer, and H. Ögelman. 1974. *Nature*, **251**:397.

Komesaroff, M. M. 1970. *Nature*, **225**:612.

Komesaroff, M. M., J. G. Ables, D. J. Cooke, P. A. Hamilton, and P. M. McCulloch. 1973. *Astrophys. Letters*, **15**:169.

Komesaroff, M. M., J. G. Ables, and P. A. Hamilton. 1971. *Astrophys. Letters*, **9**:101.

Komesaroff, M. M., P. A. Hamilton, and J. G. Ables. 1972. *Austral. J. Phys.*, **25**:759.

Kristian, J. 1970a. *Astrophys. J.* (*Letters*), **162**:L103.

———. 1970b. *Astrophys. J.* (*Letters*), **162**:L173.

Kristian, J., K. D. Clardy, and J. A. Westphal. 1976. *Astrophys. J.* (*Letters*), **206**:L143.

Kristian, J., N. Visvanathan, J. A. Westphal, and G. H. Snellen. 1970. *Astrophys. J.*, **162**:475.

Kurfess, J. D. 1971. *Astrophys. J.* (*Letters*), **168**:L39.

Kurfess, J. D., and G. H. Share. 1973. *Nature Phys. Sci.*, **244**:39.

Lamb, D. Q., and F. K. Lamb. 1976. *Astrophys. J.*, **204**:168.

Landau, L. D., and E. M. Lifshitz. 1962. *The Classical Theory of Fields*, 2nd ed., Oxford: Pergamon, p. 336.

Landstreet, J. D., and J. R. P. Angel. 1971. *Nature*, **230**:103.

Lang, K. R. 1971. *Astrophys. Letters*, **7**:175.

Large, M. I. 1971. *The Crab Nebula, I.A.U. Symposium No. 46*, Dordrecht: Reidel, p. 165.

Large, M. I., and A. E. Vaughan. 1971. *Mon. Not. Roy. Astron. Soc.*, **151**:277.

Large, M. I., A. E. Vaughan, and B. Y. Mills. 1968. *Nature*, **220**:340.

Laros, J. G., J. L. Matteson, and R. M. Pelling. 1973. *Nature Phys. Sci.*, **246**:109.

Lasker, B. M. 1976. *Astrophys. J.*, **203**:193.

Lee, L. C., and J. R. Jokipii. 1975. *Astrophys. J.*, **201**:532.

Lerche, I. 1970a. *Astrophys. J.*, **159**:229.

———. 1970b. *Astrophys. J.*, **160**:1003.

———. 1974. *Astrophys. J.*, **191**:191.

Lohsen, E. 1972. *Nature Phys. Sci.*, **236**:70.

———. 1975. *Nature*, **258**:688.

Lovelace, R. V. E., J. M. Sutton, and E. E. Salpeter. 1969. *Nature*, **222**:231.

Lucke, R., D. Yentis, H. Friedman, G. Fritz, and S. Shulman. 1976. *Astrophys. J. (Letters)*, **206**:L25.

Lynds, R., S. P. Maran, and D. E. Trumbo. 1969. *Astrophys. J. (Letters)*, **155**:L121.

Lyne, A. G. 1971. *Mon. Not. Roy. Astron. Soc.*, **153**:27P.

———. 1974. *Galactic Radio Astronomy, I.A.U. Symposium No. 60*, Dordrecht: Reidel, p. 87.

Lyne, A. G., R. T. Ritchings, and F. G. Smith. 1975. *Mon. Not. Roy. Astron. Soc.*, **171**:579.

Lyne, A. G., and D. J. Thorne. 1975. *Mon. Not. Roy. Astron. Soc.*, **172**:97.

Lyon, J. 1975. *Astrophys. J.*, **201**:168.

Macy, W. W. 1974. *Astrophys. J.*, **190**:153.

Manchester, R. N. 1971a. *Astrophys. J. Suppl. Ser.*, **23**:283.

———. 1971b. *Astrophys. J. (Letters)*, **163**:L61.

———. 1972. *Astrophys. J.*, **172**:43.

———. 1975. *Proc. Astron. Soc. Australia*, **2**:334.

Manchester, R. N., W. M. Goss, and P. A. Hamilton. 1976. *Nature*, **259**:291.

Manchester, R. N., P. A. Hamilton, W. M. Goss, and L. M. Newton. 1976. *Proc. Astron. Soc. Australia*, **3**:81.

Manchester, R. N., P. A. Hamilton, P. M. McCulloch, and J. G. Ables. 1977, in preparation.

Manchester, R. N., Tademaru, E., Taylor, J. H., and Huguenin, G. R. 1973. *Astrophys. J.*, **185**:951.

Manchester, R. N., and J. H. Taylor. 1974. *Astrophys. J. (Letters)*, **191**:L63.

Manchester, R. N., J. H. Taylor, and G. R. Huguenin. 1973. *Astrophys. J. (Letters)*, **179**:L7.

———. 1975. *Astrophys. J.*, **196**:83.

Manchester, R. N., J. H. Taylor, and Y. Y. Van. 1974. *Astrophys. J. (Letters)*, **189**:L119.

Matveyenko, L. I. 1971. *Astron. Zhurnal*, **48**:1154 (*Soviet Astron.*, **15**:918).

Matveyenko, L. I., and M. L. Meeks. 1972. *Astron. Zhurnal*, **49**:965 (*Soviet Astron.*, **16**:790).

McBreen, B., S. E. Ball, M. Campbell, K. Greisen, and D. Koch. 1973. *Astrophys. J.*, **184**:571.

McLean, A. I. O. 1973. *Mon. Not. Roy. Astron. Soc.*, **165**:133.

Mertz, L. 1974. *Astrophys. Space Sci.*, **30**:43.

Mestel, L. 1971. *Nature Phys. Sci.*, **233**:149.

——. 1973. *Astrophys. Space Sci.*, **24**:289.

Michel, F. C. 1969. *Astrophys. J.*, **158**:727.

——. 1971. *Comm. Astrophys. Space Sci.*, **3**:80.

——. 1973a. *Astrophys. J.* (*Letters*), **180**:L133.

——. 1973b. *Astrophys. J.*, **180**:207.

——. 1974. *Astrophys. J.*, **192**:713.

——. 1975. *Astrophys. J.*, **197**:193.

Migdal, A. B. 1973. *Phys. Rev. Letters*, **31**:247.

Miller, J. S. 1973. *Astrophys. J.* (*Letters*), **180**:L83.

Miller, J. S., and E. J. Wampler. 1969. *Nature*, **221**:1037.

Milne, D. K. 1970. *Austral. J. Phys.*, **23**:425.

Minkowski, R. 1942. *Astrophys. J.*, **96**: 199.

Moore, W. E., P. C. Agrawal, and G. Garmire. 1974. *Astrophys. J.* (*Letters*), **189**:L117.

Moszkowski, S. 1974. *Phys. Rev. D.*, **9**:1613.

Muncaster, G. W., and W. J. Cocke. 1972. *Astrophys. J.* (*Letters*), **178**:L13.

Mutel, R. L., J. J. Broderick, T. D. Carr, M. Lynch, M. Desch, W. W. Warnock, and W. K. Klemperer. 1974. *Astrophys. J.*, **193**:279.

Nelson, J., R. Hills, D. Cudaback, and J. Wampler. 1970. *Astrophys. J.* (*Letters*), **161**:L235.

O'Connell, D. J. K., ed. 1958. "Stellar Populations," *Ric. Astr. Specola Vaticana*, **5**.

O'Dell, C. R. 1962. *Astrophys. J.*, **136**:809.

O'Dell, S. L., and L. Satori. 1970. *Astrophys. J.* (*Letters*), **161**:L63.

Ögelman, H., C. E. Fitchel, D. A. Kniffen, and D. J. Thompson. 1976. *Astrophys. J.*, **209**:584.

Okamoto, I. 1974. *Mon. Not. Roy. Astron. Soc.*, **167**:457.

Oke, J. B. 1969. *Astrophys. J.* (*Letters*), **156**:L49.

Oort, J. H. 1965. "Stellar Dynamics," in *Galactic Structure*, A. Blaauw and M. Schmidt, eds., University of Chicago Press, p. 455.

Oppenheimer, J. R., and G. Volkoff. 1939. *Phys. Rev.*, **55**:374.

Oster, L. 1975. *Astrophys. J.*, **196**:571.

Oster, L., and W. Sieber. 1976. *Astrophys. J.*, **203**:233.

Ostriker, J. P. 1968, *Nature*, **217**:1127.

Ostriker, J. P., and J. E. Gunn. 1969. *Astrophys. J.*, **157**:1395.

Ostriker, J. P., D. O. Richstone, and T. X. Thuan. 1974. *Astrophys. J.* (*Letters*), **188**:L87.

Pacini, F. 1967. *Nature*, **216**:567.

——. 1968. *Nature*, **221**:454.

——. 1971. *Astrophys. J.* (*Letters*), **163**:L17.

Paczynski, B. E. 1971. *Ann Rev. Astron. Astrophys.*, **9**:183.

———. 1973. *Late Stages of Stellar Evolution*, I.A.U. Symposium No. 66, Dordrecht: Reidel, p. 62.

Pandharipande, V. R., D. Pines, and R. A. Smith. 1976. *Astrophys. J.*, **208**:550.

Papaliolios, C., N. P. Carleton, and P. Horowitz. 1970. *Nature*, **228**:445.

Papaliolios, C., and P. Horowitz. 1973. *Astrophys. J.*, **183**:233.

Parker, E. A. 1968. *Mon. Not. Roy. Astron. Soc.*, **138**:407.

Piddington, J. H. 1957. *Austral. J. Phys.*, **10**:530.

Pines, D., and J. Shaham. 1972. *Nature Phys. Sci.*, **235**:43.

Pines, D., J. Shaham, and M. Ruderman. 1972. *Nature Phys. Sci.*, **237**:83.

Plavec, M. 1968. *Advances Astron. Astrophys.*, **6**:201.

Pravdo, S. H., R. H. Becker, E. A. Bolt, S. S. Holt, R. E. Rothschild, P. J. Serlemitsos, and J. H. Swank. 1976. *Astrophys. J. (Letters)*, **208**:L67.

Prentice, A. J. R., and D. ter Haar. 1969, *Mon. Not. Roy. Astron. Soc.*, **146**:423.

Radhakrishnan, V., and D. J. Cooke. 1969. *Astrophys. Letters*, **3**:225.

Radhakrishnan, V., D. J. Cooke, M. M. Komesaroff, and D. Morris. 1969. *Nature*, **221**:443.

Radhakrishnan, V., V. M. Goss, J. D. Murray, and J. W. Brooks. 1972. *Astrophys. J. Suppl. Ser.*, **24**:49.

Radhakrishnan, V., and R. N. Manchester. 1969. *Nature*, **222**:228.

Rankin, J. M., D. B. Campbell, and D. C. Backer. 1974. *Astrophys. J.*, **188**:609.

Rankin, J. M., J. M. Comella, H. D. Craft, D. W. Richards, D. B. Campbell, and C. C. Counselman. 1970. *Astrophys. J.*, **162**:707.

Rankin, J. M., and C. C. Counselman. 1973. *Astrophys. J.*, **181**:875.

Rankin, J. M., R. R. Payne, and D. B. Campbell. 1974. *Astrophys. J. (Letters)*, **193**:L71.

Rappaport, S., H. Bradt, R. Doxsey, A. Levine, and G. Spada. 1974. *Nature*, **251**:471.

Rappaport, S., H. Bradt, and W. Mayer 1971. *Nature Phys. Sci.*, **229**:40.

Rappaport, S., P. C. Joss, and J. E. McClintock. 1976. *Astrophys. J.*, **206**:L103.

Readhead, A. C. S., and P. J. Duffet-Smith. 1975. *Astron. Astrophys.*, **42**:151.

Rees, M. J. 1971. *Nature Phys. Sci.*, **230**:55.

Rees, M. J., and J. E. Gunn. 1974. *Mon. Not. Roy. Astron. Soc.*, **167**:1.

Reichley, P. E., and G. S. Downs. 1969. *Nature*, **222**:229.

———. 1971. *Nature Phys. Sci.*, **234**:48.

Rhoades, C. E., and R. Ruffini. 1974. *Phys. Rev. Letters*, **32**:324.

Richards, D. W., and J. M. Comella. 1969. *Nature*, **222**:551.

Ricker, G. R., A. Scheepmaker, S. G. Ryckman, J. E. Ballintine, J. P. Doty, P. M. Downey, and W. H. G. Lewin. 1975. *Astrophys. J. (Letters)*, **197**:L83.

Rickett, B. J. 1969. *Nature*, **221**:158.

———. 1975. *Astrophys. J.*, **197**:185.

Rickett, B. J., T. H. Hankins, and J. M. Cordes. 1975. *Astrophys. J.*, **201**:425.

Rickett, B. J., and K. R. Lang. 1973. *Astrophys. J.*, **185**:945.

Ritchings, R. T. 1976. *Mon. Not. Roy. Astron. Soc.*, **176**:249.

Ritchings, R. T., and A. G. Lyne. 1975. *Nature*, **257**:293.

Roberts, D. H., A. R. Masters, and W. D. Arnett. 1976. *Astrophys. J.*, **203**:196.

Roberts, D. H., and P. A. Sturrock. 1972a. *Astrophys. J.*, **172**:435.

――. 1972b. *Astrophys. J. (Letters)*, **173**:L33.

――. 1973. *Astrophys. J.*, **181**:161.

Roberts, J. A., and D. W. Richards. 1971. *Nature Phys. Sci.*, **231**:25.

Robinson, B. J., B. F. C. Cooper, F. F. Gardner, R. Wielebinski, and T. L. Landecker. 1968. *Nature*, **218**:1143.

Ruderman, M. A. 1971. *Phys. Rev. Letters*, **27**:1306.

――. 1972. *Ann. Rev. Astron Astrophys.*, **10**:427.

――. 1976a. *Astrophys. J.*, **203**:213.

――. 1976b. *Astrophys. J.*, **203**:206.

Ruderman, M. A., and P. G. Sutherland. 1973. *Nature Phys. Sci.*, **246**:93.

――. 1974. *Astrophys. J.*, **190**:137.

――. 1975. *Astrophys. J.*, **196**:51.

Rumsey, V. H. 1975, *Radio Science*, **10**:107.

Scargle, J. D. 1969. *Astrophys. J.*, **156**:401.

Scargle, J. D., and F. Pacini. 1971. *Nature Phys. Sci.*, **232**:144.

Scharlemann, E. T. 1974. *Astrophys. J.*, **193**:217.

Scharlemann, E. T., and R. V. Wagoner. 1973. *Astrophys. J.*, **182**:951.

Scheuer, P. A. G. 1968. *Nature*, **218**:920.

Schönhardt, R. E. 1973. *Nature Phys. Sci.*, **243**:62.

Schönhardt, R. E., and W. Sieber. 1973. *Astrophys. Letters*, **14**:61.

Seiradakis, J. H. 1975. *A Pulsar Survey*, thesis, University of Manchester.

Shaver, P. A., A. Pedlar, and R. D. Davies. 1976. *Mon. Not. Roy. Astron. Soc.*, **177**:45.

Shklovskii, I. S. 1969. *Astron. Zhurnal*, **46**:715, and *Soviet Astron.*, **13**:562.

――. 1970. *Astrophys. J. (Letters)*, **159**:L77.

Sieber, W. 1973. *Astron. Astrophys.*, **28**:237.

Sieber, W., and L. Oster. 1975. *Astron. Astrophys.*, **38**:325.

Sieber, W., R. Reinecke, and R. Wielebinski. 1975. *Astron. Astrophys.*, **38**:169.

Slee, O. B., J. G. Ables, R. A. Batchelor, S. Krishna-Mohan, V. R. Venugopal, and G. Swarup. 1974. *Mon. Not. Roy. Astron. Soc.*, **167**:31.

Smarr, L. L., and R. Blandford. 1976. *Astrophys. J.*, **207**:574.

Smith, F. G. 1970. *Mon. Not. Roy. Astron. Soc.*, **149**:1.

――. 1973. *Nature*, **243**:207.

Smith, H. E., B. Margon, and S. Conti. 1973. *Astrophys. J. (Letters)*, **179**:L125.

Spitzer, L., and E. B. Jenkins. 1975. *Ann. Rev. Astron. Astrophys.*, **13**:133.

Staelin, D. H. 1969. *Proc. Inst. Elec. Electron. Engrs.*, **57**:724.

Staelin, D. H., and E. C. Reifenstein. 1968. *Science*, **162**:1481.

Sturrock, P. A. 1971. *Astrophys. J.*, **164**:529.

Sturrock, P. A., V. Petrosian, and J. S. Turk. 1975. *Astrophys. J.*, **196**:73.

Sutton, J. M. 1971. *Mon. Not. Roy. Astron. Soc.*, **155**:51.

Sutton, J. M., D. H. Staelin, R. M. Price, and R. Weimer. 1970. *Astrophys. J.* (*Letters*), **159**:L89.

Tademaru, E. 1971. *Astrophys. Space Sci.*, **12**:193.

———. 1973. *Astrophys. J.*, **183**:625.

———. 1974. *Astrophys. Space Sci.*, **30**:179.

Tademaru, E., and E. R. Harrison. 1975. *Nature*, **254**:676.

Takemori, M. T., and R. A. Guyer. 1975. *Phys. Rev. D.* **11**:2696.

Tammann, G. A. 1974. "Statistics of Supernovae." In *Supernovae and Supernovae Remnants*, C. B. Cosmovici, ed., Dordrecht: Reidel, p. 155.

Tanenbaum, B. S., G. A. Zeissig, and F. D. Drake. 1968. *Science*, **160**:760.

Taylor, J. H. 1974. *Astron. Astrophys. Suppl. Ser.*, **15**:367.

Taylor, J. H., and G. R. Huguenin. 1971. *Astrophys. J.*, **167**:273.

Taylor, J. H., R. A. Hulse, L. A. Fowler, G. E. Gullahorn, and J. M. Rankin. 1976. *Astrophys. J.* (*Letters*), **206**:L53.

Taylor, J. H. and R. N. Manchester. 1975. *Astron. J.*, **80**:794.

———. 1977. *Astrophys. J.*, **215** (August 1).

Taylor, J. H., R. N. Manchester, and G. R. Huguenin. 1975. *Astrophys. J.*, **195**:513.

Thomas, R. M., and R. Rothenflug. 1974. *Nature*, **249**:812.

Thomas, R. M. 1975. *Proc. Astron. Soc. Austral.*, **2**:325.

Thompson, D. J. 1975. *Astrophys. J.* (*Letters*), **201**:L117.

Thompson, D. J., C. E. Fichtel, D. A. Kniffen, and H. B. Ögelman. 1975. *Astrophys. J.* (*Letters*), **200**:L79.

Trimble, V. 1968. *Astron. J.*, **73**:535.

———. 1971. *The Crab Nebula, I.A.U. Symposium No. 46*, Dordrecht: Reidel, p. 12.

Tsuruta, S., V. Canuto, J. Lodenquai, and M. Ruderman. 1972. *Astrophys. J.*, **176**:739.

Tuohy, I. R. 1976. *Mon. Not. Roy. Astron. Soc.*, **174**:45P.

Uscinski, B. J. 1975. *Mon. Not. Roy. Astron. Soc.*, **172**:117.

Vandenberg, N. R., T. A. Clark, W. C. Erickson, G. M. Resch, J. J. Broderick, R. R. Payne, S. H. Knowles, and A. B. Youmans. 1973. *Astrophys. J.* (*Letters*), **180**:L27.

Van den Bergh, S., A. P. Marscher, and Y. Terzian. 1973. *Astrophys. J. Suppl.*, **26**:19.

Van den Heuvel, E. P. J. 1975a. *Astrophys. J.* (*Letters*), **198**:L109.

———. 1975b. Enrico Fermi Summer School on Physics and Astrophysics of Neutron Stars and Black Holes, Varenna, Italy.

Van Paradijs, J. A., G. Hammerschlag-Hensberge, E. P. J. van den Heuvel, R. J. Takens, E. J. Zuiderwijk, and C. De Loore. 1976. *Nature*, **259**:547.

Villa, G., C. G. Page, M. J. L. Turner, B. A. Cooke, M. J. Ricketts, K. A. Pounds, and D. J. Adams. 1976. *Mon. Not. Roy. Astron. Soc.*, **176**:609.

Virtamo, J., and P. Jauho. 1973. *Astrophys. J.*, **182**:935.

Wagoner, R. V. 1975. *Astrophys. J.* (*Letters*), **196**:L63.

Wallace, P. T., B. A. Peterson, P. G. Murdin, I. J. Danziger, R. N. Manchester, A. G. Lyne, W. M. Goss, F. G. Smith, M. J. Disney, K. F. Hartley, D. H. P. Jones, and G. W. Wellgate. 1977. *Nature*, **266**:692.

Wampler, E. J. 1972. "Optical Observations of Pulsars," in *The Physics of Pulsars*, A. M. Lenchek, ed., New York: Gordon and Breach, p. 21.

Warner, B., R. E. Nather, and M. Macfarlane. 1969. *Nature*, **222**:233.

Webbink, R. F. 1975. *Astron. Astrophys.*, **41**:1.

Weiler, K. W. 1975. *Nature*, **253**:24.

Weisskopf, M. C., G. G. Cohen, H. L. Kestenbaum, K. S. Long, R. Novick, and R. S. Wolff. 1976. *Astrophys. J.* (*Letters*), **208**:L125.

Wentzel, D. G. 1974. *Ann. Rev. Astron. Astrophys.*, **12**:71.

Wheeler, J. A. 1966. *Ann. Rev. Astron. Astrophys.*, **4**:393.

Wheeler, J. C., M. Lecar, and C. F. McKee. 1975. *Astrophys. J.*, **200**:145.

Will, C. M. 1976. *Astrophys. J.*, **205**:861.

Williams, D. R. W., W. J. Welch, and D. D. Thornton. 1965. *Publ. Astron. Soc. Pacific*, **77**:178.

Williamson, I. P. 1974. *Mon. Not. Roy. Astron. Soc.*, **166**:499.

Wilson, A. S. 1972a. *Mon. Not. Roy. Astron. Soc.*, **157**:229.

———. 1972b. *Mon. Not. Roy. Astron. Soc.*, **160**:373.

———. 1974. *Mon. Not. Roy. Astron. Soc.*, **166**:617.

Winkler, P. F., and G. W. Clark. 1974. *Astrophys. J.* (*Letters*), **191**:L67.

Wolff, R. S., H. L. Kestenbaum, W. Ku, and R. Novick. 1975. *Astrophys. J.* (*Letters*), **202**:L77.

Wolff, S. C., and N. D. Morrison. 1974. *Astrophys. J.*, **187**:69.

Wolszczan, A., K. H. Hesse, and W. Sieber. 1974. *Astron. Astrophys.*, **37**:285.

Woltjer, L. 1957. *Bull. Astron. Inst. Neth.*, **13**:302.

Zheleznyakov, V. V. 1970. *Izv. VUZ Radiofiz.*, **13**:1842.

———. 1971. *Astrophys. Space Sci.*, **13**:87.

Zheleznyakov, V. V., and V. E. Shaposhnikov. 1972. *Astrophys. Space Sci.*, **18**:141.

Zimmermann, H. U. 1974. *Astron. Astrophys.*, **34**:305.

Indexes

Name Index

Subject Index